Sturge-Weber Syndrome

Sturge-Weber Syndrome

Edited by
John B. Bodensteiner, M.D. and E. Steve Roach, M.D.

SWF

Copyright © 2010

Published by:
The Sturge-Weber Foundation
P.O. Box 418
Mt. Freedom, NJ 07970
(800) 627-5482
(973) 895-4445
(973) 895-4846 Fax
www.sturge-weber.org

Version History
First Edition published June 1999
ISBN-10: 0967048400
ISBN-13: 978-0967048406

Second Edition published October 2010
ISBN-10: 0983060606
ISBN-13: 9780983060604

Layout and Design by Tom Heffron

Printed in the United States of America

Contents

We dedicate this book to Donna Bodensteiner and Lisa Roach –
in recognition of their patience and support.

Contributors

Bálint Alkonyi, M.D.
Research Associate
PET Center, Children's Hospital of Michigan
Detroit, Michigan

Karen L. Ball, President and CEO
The Sturge-Weber Foundation
Mt. Freedom, New Jersey 07970-0418

John B. Bodensteiner, M.D.
William Pilcher Chair of Pediatric Neurology
St. Joseph's Children's Health Center
The Barrow Neurologic Institute Phoenix
Phoenix, Arizona

Joanna M. Burch, M.D.
Associate Professor of Dermatology and Pediatrics
University of Colorado School of Medicine
Director, Pediatric Dermatology Clinic at Denver Children's Hospital
Denver, Colorado

Lynn Chapieski, Ph.D.
Associate Professor of Pediatrics
Pediatric Neuropsychology
Baylor College of Medicine
Texas Children's Hospital

Harry T. Chugani, M.D.
Professor and Division Chief
Pediatric Neurology and PET Center
Children's Hospital of Michigan
Detroit, Michigan

Anne M. Comi, M.D.
Associate Professor of Neurology and Pediatrics
Kennedy Krieger Institute
Johns Hopkins School of Medicine
Baltimore, Maryland

Sharon F. Freedman, M.D.
Professor of Ophthalmology and Pediatrics
Duke University Eye Center
Durham, North Carolina

Ronald T. Grondin, M.D., M.Sc., F.R.C.S.C.
Assistant Professor of Neurosurgery
Ohio State University College of Medicine
Attending Physician, Nationwide Children's Hospital
Columbus, Ohio

Geoffrey Heyer, M.D.
Division of Child Neurology
The Ohio State University College of Medicine
Attending Physician, Nationwide Children's Hospital
Columbus, Ohio

Csaba Juhász, M.D., Ph.D.
Associate Professor of Pediatrics and Neurology
Wayne State University School of Medicine
PET Center, Children's Hospital of Michigan
Detroit, Michigan

Joseph G. Morelli, M.D.
Professor of Dermatology and Pediatrics
University of Colorado Denver School of Medicine
Head, Section of Pediatric Dermatology, Denver Children's Hospital
Denver, Colorado

E. Steve Roach, M.D.
Professor of Pediatrics and Neurology
The Ohio State University College of Medicine
Attending Physician, Nationwide Children's Hospital
Columbus, Ohio

Tammy L. Yanovitch, M.D., M.HSc.
Assistant Professor of Pediatric Ophthalmology
Duke University Eye Center
Durham, North Carolina

Foreword

Sturge-Weber syndrome remains one of the most enigmatic of the neurocutaneous disorders in terms of its etiology, pathogenesis, and lack of Mendelian genetic transmission. Nevertheless, since the first edition of this book in 1999, our understanding of the clinical presentation, course, and treatment of the many neurological and non-neurological complications have improved, and these advances are well defined in this updated edition. The monograph is comprehensive in its scope and farsighted in identifying aspects that require further study for a more insightful comprehension. It is well illustrated and contains a wealth of both classical and recent references that provide a valuable and time-saving resource for readers.

The pathophysiology of Sturge-Weber syndrome is still poorly understood, particularly in regard to the contribution of genetic mutations or the aberrant expression of genes of angiogenesis; the pathogenesis was better defined during the past decade. The correspondence of the distribution of cutaneous angiomata to the three branches of the trigeminal nerve are strongly suggestive of a neural crest defect, because nerve sheaths, blood vessels, and cranial meninges are all neural crest derivatives, mostly arising in the embryonic mesencephalic and rhombencephalic neural crest in particular. Vascular nevi in segments below rhombomeres r1-r3 (metencephalon, later the pons), as expressed in a minority of patients with "port wine stains" on the neck or upper thorax in addition to the face, are rhombencephalic neural crest territories as well. Even if a gene is discovered to elucidate this phenomenon, it still might need further elaboration to explain why it is expressed asymmetrically or unilaterally in most patients. The authors also clearly address the issue of the individual with a facial cutaneous vascular nevus but not a meningeal or cerebral lesion and the distinction of this condition from Sturge-Weber syndrome.

Chapter 4 by Comi, Roach, and Bodensteiner reviews the various clinical neurological findings, some neuroimaging features (later amplified in Chapter 6, by Juhász, Alkonyi, and Chugani, which provides an in-depth description and analysis of the MRI and functional imaging features of the syndrome), EEG and other neurophysiological features, and a well focused discussion of epilepsy in this disorder. Epidemiological data are provided to define the frequency of various clinical aspects, such as seizures and headache. Moreover, despite the absence of an entire chapter dedicated to neuropathology, I was pleased to see that these authors, all of whom are distinguished clinical pediatric neurologists, not only recognize the importance of neuropathological findings and include a subsection identified with a subtitle, "Neuropathology of Sturge-Weber Syndrome," but they use it, with considerable success, to explain many of the neuroimaging features that can be fully understood only at a microscopic level.

Indeed they go beyond the standard histopathological correlations and review the recent molecular neuropathology, including such items as vascular endothelial growth factor (VEGF) and its receptors on meningeal blood vessels, including possible implications for future therapeutic approaches. They balance this enthusiasm by prudent caution in recognizing risks of interfering with the dynamics of vascular remodeling, in case these factors actually might be contributing a compensatory mechanism in an attempt to minimize the lesion. Such thoughtfulness will surely provide hypotheses that can be tested in animals and later extrapolated to the human. Finally, current treatment options, including seizure control, anti-migrainous therapy for the headaches, and anti-platelet medications are each carefully considered and discussed.

The breadth of the treatise is apparent even from the Contents, which identifies chapters dealing with ophthalmological, psychological, and neurosurgical aspects, and also congenital venous abnormalities in other organ systems of the body in this disorder. The final chapter about the Sturge-Weber Foundation will be useful to clinicians who wish to refer their patients to a site where they can obtain more information and also learn about research activities related to this disorder.

It is an honor for me to have been invited to write a foreword to such an excellent monograph. It can serve as a resource to any physician managing patients with Sturge-Weber syndrome.

Harvey B. Sarnat, M.S., M.D., F.R.C.P.C.
Professor of Paediatrics, Pathology (Neuropathology) and
    Clinical Neurosciences
University of Calgary Faculty of Medicine and Alberta Children's Hospital
Calgary, Alberta, Canada
harvey.sarnat@albertahealthservices.ca

Preface to the First Edition

SINCE THE PUBLICATION OF *The Sturge-Weber Syndrome* almost forty years ago by G. L. Alexander and R. M. Norman, major advances in neuroimaging have occurred, and new treatment methods have been developed for some of the syndrome's most serious complications. Contrast-enhanced computed cranial tomography and magnetic resonance imaging have supplanted routine roentgenography, cerebral angiography, and nuclide imaging as the imaging techniques used to demonstrate the nature and extent of the intracranial vascular lesion of Sturge-Weber syndrome. Functional neuroimaging with PET and SPECT have provided insight into how the vascular lesions promote the neurologic complications of Sturge-Weber syndrome. More effective antiepileptic drugs and increasing experience with surgical procedures for epilepsy have improved our ability to treat the seizures that plague most patients with Sturge-Weber syndrome. Treatment methods for glaucoma have improved significantly, and new laser techniques sometimes enable the cutaneous nevus of Sturge-Weber syndrome to be obliterated or at least minimized.

Our understanding of the clinical characteristics of Sturge-Weber syndrome has also improved over the years. Once considered a uniformly devastating neurologic condition, the highly variable nature of Sturge-Weber syndrome is now known to include many individuals without weakness, people whose epilepsy is controllable with medication, and those with relatively normal cognitive function. We now recognize that some patients have episodic neurologic dysfunction as a result of a vascular phenomenon, and we better appreciate the deleterious effects of epileptic seizures on the developing brain.

For all our progress in the diagnosis and treatment of Sturge-Weber syndrome, our understanding of the disorder's pathophysiology remains comparatively limited. The embryology of the cerebrovascular system is better understood now, and the process of angiogenesis is beginning to be unraveled. Nevertheless, there are many unanswered questions about Sturge-Weber syndrome.

Sturge-Weber syndrome is an enigmatic disorder, seldom difficult to diagnose but often difficult to treat. It is our hope that this book will consolidate what is known about the Sturge-Weber syndrome in a single source, making it easier to apply this new information to the care of individual patients and perhaps stimulating research on some of the remaining questions about the syndrome.

John B. Bodensteiner, M.D.

E. Steve Roach, M.D.

Preface to the Second Edition

IN THE 11 YEARS SINCE THE FIRST EDITION of this book, the reasons for writing the book have not changed much. Our desire to update and consolidate the available information regarding the Sturge-Weber syndrome remains the driving motivation. Many of the new technologies that are applied to the diagnosis and management of Sturge-Weber syndrome existed eleven years ago. As with most technological advances in medicine, however, we have gained considerable experience with these technologies, and know with more confidence how to properly apply the advances to the problem of Sturge-Weber syndrome. To use neuroimaging as an example, we have benefitted from newer generations of scanners, and new variations on these technologies, such as tractography, have been developed. The availability and resolution of PET scanning has improved dramatically in this time. As each new technology has evolved, we grapple anew with delineating the limits of the technology and how to apply the technique in the most advantageous way for the benefit of our patients with Sturge-Weber syndrome. It is in this spirit that we have revised and updated *Sturge-Weber Syndrome*, resulting in this second edition.

We are grateful to for the ongoing support for this project by Karen Ball and others at the Sturge-Weber Foundation for their ongoing support for the revision and for providing the wonderful photographs on the back cover. We are indebted to Ms. Gail White for editorial assistance. Sydney Schochet, M.D. has passed away since the first edition was published, but he provided several photographs used in both editions. We gratefully acknowledge Drs. Csaba Juhász and Harry Chugani for providing the beautiful cover pictures.

John B. Bodensteiner, M.D.

E. Steve Roach, M.D.

Chapter 1

Overview of Sturge-Weber Syndrome

John B. Bodensteiner, M. D.

E. Steve Roach, M.D.

THE CUTANEOUS HALLMARK OF STURGE-WEBER SYNDROME, the facial angioma or port-wine stain, has been recognized since antiquity. In those early times, this "mother's mark" was often attributed to an adverse event, such as fright, experienced by the mother during gestation, or taken as a sign from a higher being. Individuals with neurological deficits accompanying the stain were sometimes considered cursed. It was not until the latter part of the nineteenth century that the range of manifestations of the condition we call Sturge-Weber syndrome were defined medically.

EARLY DESCRIPTIONS OF THE SYNDROME

The initial medical report of the association of facial angiomatous nevi with other features is generally attributed to Schirmer, who in 1860 described the occurrence of unilateral buphthalmos in a patient with bilateral facial angiomatosis.[1] The occurrence of choroidal angiomatosis in this condition was described several years later.[2]

The occurrence of neurological impairment in an individual with a facial angiomatous nevus was first clearly delineated by Sturge, who in 1879 described a young girl with port-wine-colored angiomatosis of the face, neck, and trunk, ipsilateral buphthalmos, and contralateral seizures and weakness.[3] Sturge's name has since been linked to this condition because he clearly described the chief clinical manifestations of the disease and also correctly surmised that the hemiparesis and seizures resulted from a lesion similar to the facial nevus of the ipsilateral cerebral hemisphere. Sturge's hypothesis was confirmed by Kalischer, who in 1897 described the intracranial pathology of the condition.[4] In 1922, Weber first reported the radiographic features of the syndrome but did not describe the intracranial calcifications that typify this condition.[5] In 1923, Dimitri described the double contoured "tram track" calcifications seen on skull radiographs (**Figure 1-1**), but it was not until a decade later that Krabbe identified the source of these lines as layered calcium deposits within the cerebral cortex beneath the meningeal angioma.[6,7] Our understanding of the pathogenesis and clinical manifestations of Sturge-Weber syndrome has advanced with every new neuroimaging technique-cerebral angiography, nuclide imaging, computed cranial tomography, magnetic resonance imaging, and functional imaging. Computerized imaging techniques have largely replaced the plain skull radiographs in the radiographic evaluation of these patients, so the characteristic tram track calcifications are now seldom seen.

Although the eponym Sturge-Weber syndrome fails to acknowledge several important early observers, it remains the most common name for this disorder, largely because of convention and widespread recognition of the term. There is periodic debate about how best to define and characterize the syndrome, but there seems to be no good reason to change its name until we have a better

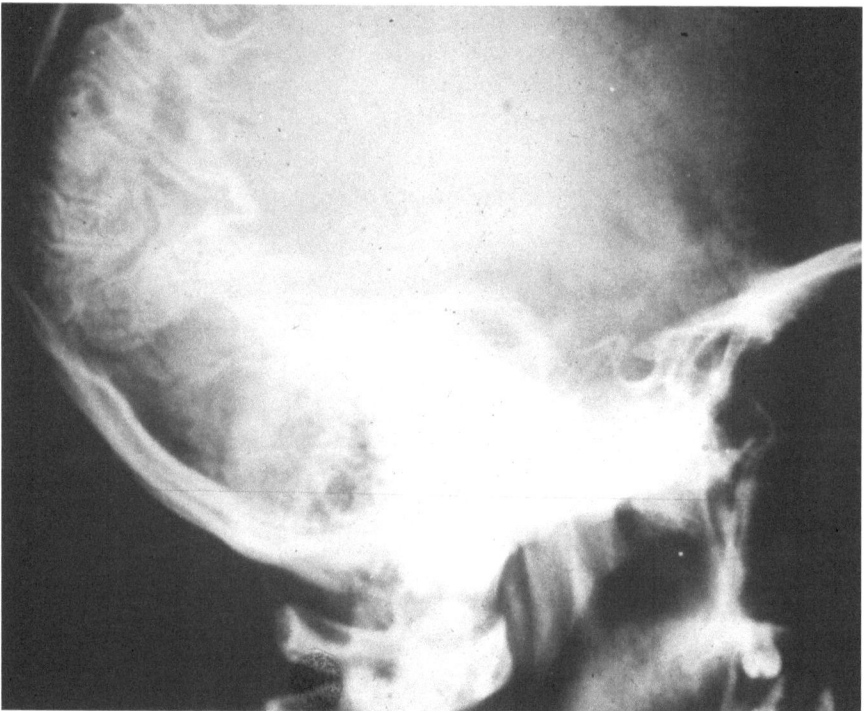

Figure 1-1. Lateral radiograph of the skull showing the characteristic "tram track" calcifications first described by Dimitri in 1923. *S. S. Schochet*

understanding of both the molecular mechanisms that cause angiomatosis and the factors that determine the location and extent of the abnormal vessels. Then, a more descriptive, definitive name may be warranted.

DEFINITION OF THE CONDITION

The two most constant features of Sturge-Weber syndrome (encephalofacial angiomatosis) are a facial port-wine stain (**Figure 1-2**) and clinical or radiographic evidence of ipsilateral brain dysfunction (**Figure 1-3**). Some individuals have glaucoma, dental abnormalities, and skeletal lesions. Most of the patients develop epilepsy, while mental retardation is common but by no means universal. Less common neurological manifestations include contralateral hemiparesis, hemiatrophy, and homonymous hemianopia.[8]

The array of clinical features is variable and the severity differs greatly from one individual to another. At one end of the spectrum are patients with facial port-wine stains and seizures as the only manifestations of the meningeal angiomatosis. Such patients may have normal intelligence, no ocular involvement, and no identifiable neurologic dysfunction aside from their seizures.[9] In contrast, some patients have severe and sometimes progressive cognitive and neurological

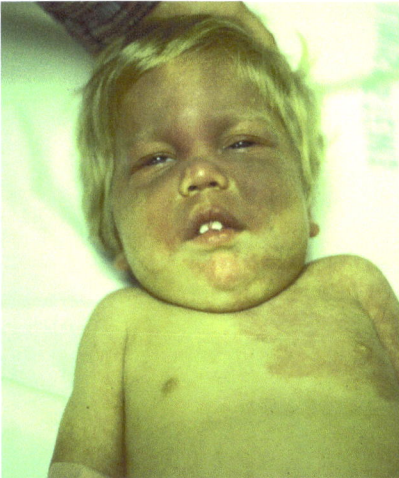

Figure 1-2A. Photograph of an infant with extensive bilateral facial port-wine stain extending over the left shoulder. *Reproduced with permission from Seminars in Neurology* 1988;8:83-96

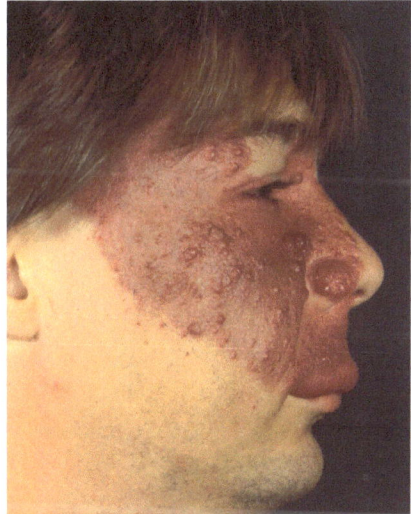

Figure 1-2B. Port-wine stain in an adult with Sturge-Weber syndrome showing the thickening and nodularity that may occur in the cutaneous lesions.

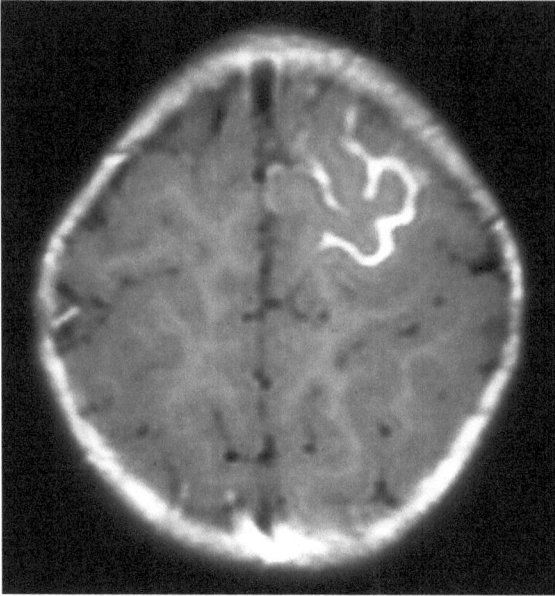

Figure 1-3. Magnetic resonance image with gadolinium contrast from a child with Sturge-Weber syndrome shows abnormal venous channels with enhancement of the angioma in the left frontal lobe.

deficits, intractable seizures, increased intracranial pressure, intractable glaucoma with visual loss, and dental problems.[10]

The name Klippel-Trenaunay-Weber syndrome has been assigned to the condition of those children with extensive somatic involvement of the limbs and trunk, particularly with enlargement of the affected structures. From a clinical standpoint, however, involvement of the upper face would indicate that these individuals risk the same neurological problems as patients with Sturge-Weber syndrome.

The diagnosis of Sturge-Weber syndrome is seldom difficult, but it is hard to predict the eventual functional outcome or to guess when complications will occur. While treatment of many of these complications has improved in the last few years, effective therapy with preservation of function often remains distressingly difficult. Unlike several other neurocutaneous syndromes, Sturge-Weber syndrome is not inherited as a Mendelian trait, although it could result from somatic mosaicism of one or more genes that direct vascular development. The syndrome presents in all races and with equal frequency in both sexes.[11-13]

CLINICAL FEATURES
Cutaneous manifestations

Capillary angiomas are relatively common, but most such patients do not have the neurologic complications of Sturge-Weber syndrome. The cutaneous capillary angioma of Sturge-Weber syndrome (the port-wine stain) is classically found on the forehead and upper eyelid (**Figure 1-2**) ipsilateral to the intracranial leptomeningeal angioma.[3] Extension of the lesion onto both sides of the face and onto the trunk and extremities is also common,[9] but the risk of having a leptomeningeal angioma is small when only the trunk or lower face is affected.[14] Even when a typical lesion of the upper face is present, leptomeningeal angiomatosis occurs only 3% to 20% of the time.[9,14-16] A facial angioma is usually obvious at birth. A few patients have the typical neurological features of Sturge-Weber syndrome without a nevus.[9,17-19]

When the facial lesion is unilateral, the leptomeningeal angioma is typically ipsilateral to the nevus. However, bilateral leptomeningeal angiomas are not unusual in individuals with a unilateral facial lesion and are common in those with a bilateral facial lesion.[20] The extent of the facial angioma correlates poorly with the severity or extensiveness of leptomeningeal involvement, although children with extensive bilateral facial involvement are more likely to have bilateral brain lesions, and consequently they have a greater risk of neurological impairment and earlier onset seizures.[9,16,21]

The port-wine lesions are characterized by dermal and subcutaneous collections of dilated capillary-like vessels. Despite the significance traditionally accorded it, the crucial overlap of the angioma with the area innervated by the first division of the trigeminal nerve may be a mere coincidence. More important, the frequently involved facial and leptomeningeal areas share a common embryological origin.[22]

Some of the more troublesome cutaneous lesions involve the lips, gingiva, tongue, palate, pharynx, and larynx, which with time may thicken and cause progressive distortion and disfigurement.[21] The port-wine stain may become more prominent and darker with time.[23] While the lesion does not usually widen, the nevus often becomes thicker and nodular over time, presumably due to dilatation and ectasia of the lesion's vessels (**Figure 1-2B**). This tendency for the port wine stain to thicken and develop nodularity is one reason for early treatment of the nevus. Treatment options for port-wine stains, primarily laser obliteration of the abnormal skin vessels, is discussed in Chapter 2.

Ophthalmological Features

Glaucoma is the most serious ophthalmological complication of Sturge-Weber syndrome. The pathogenesis of the increased intraocular pressure is not always clear and may be associated with outflow obstruction by a malformation of the anterior chamber of the eye, hypersecretion from an associated angioma of the choroid, elevated venous pressure associated with an episcleral hemangioma, or a combination of these possibilities.[24-27] Glaucoma occurs in 30% to 70% of patients with Sturge-Weber syndrome.[28,29] Buphthalmos (congenital enlargement of the globe), is associated with congenital glaucoma and occurs in about half of Sturge-Weber patients.[28] The risk of developing glaucoma is highest in the first decade of a patient's life, but some patients maintain normal intraocular pressure until adulthood, necessitating continued vigilance even among patients whose intraocular pressure is initially normal.[29]

Other ocular and ophthalmological complications of Sturge-Weber syndrome include angiomas of the choroid, episcleral and conjunctival lesions, heterochromia of the iris, and visual field defects. Angioma of the choroid occurs somewhat less frequently than does glaucoma, indicating that this anomaly is not the only cause of increased intraocular pressure. Between 30% and 40% of patients with Sturge-Weber syndrome have angiomatous involvement of the choroid.[30,31]

Episcleral and conjunctival lesions may occur in the form of anomalous vessels or true angiomas. These abnormal vessels may contribute to the development of glaucoma by impairing drainage of the aqueous. In one study, episcleral vascular lesions were found in all Sturge-Weber syndrome patients with glaucoma, suggesting that elevated venous pressure may be a major factor in the development of increased intraocular pressure.[25] With heterochromia of the iris, the more deeply pigmented iris is typically ipsilateral to the facial angioma. Visual field defects are common among patients whose occipital cortex is affected.

Pharmacologic agents for glaucoma are useful for some patients, but many others still require surgery. Innovative procedures to control intraocular pressure, prevent intra-ocular hemorrhage, and stabilize retinal detachment have also been added to the list of treatment options.[32-35] Treatment of the ocular complications of Sturge-Weber syndrome is discussed in Chapter 3.

Neurological Features

The principal neurologic features of patients with Sturge-Weber syndrome are epilepsy, focal neurological deficits, and mental retardation (see Chapter 4). Seizures are frequently the first neurological symptom to appear and often start in temporal association with the hemiparesis. The seizures are usually focal at first but may become secondarily generalized tonic-clonic seizures. Infantile spasms or myoclonic or atonic seizures occur less frequently.[36-38]

Seizures eventually develop in 72% to 80% of those individuals with Sturge-Weber syndrome who have unilateral leptomeningeal involvement and in up to 93% of patients with bilateral involvement.[21,39] Although seizures may develop in patients at any time, 75% of cases begin by one year of age, 86% by age two, and 95% by age five.[29] Seizure control with anticonvulsant medication has an unpredictable rate of success. Some patients experience clusters of intense seizure activity interspersed with long seizure-free intervals even without medication, whereas others have frequent or prolonged seizures despite high doses of multiple medications.[36,40] Seizure onset before one year of age increases the likelihood of future cognitive impairment, as does the occurrence of refractory seizures.[29] As a rule, individuals who have frequent seizures, particularly at a very young age, have a greater likelihood of mental retardation.[41] Among patients destined to have refractory seizures, complete lesion resection may improve cognitive outcome if it results in complete seizure control (see Chapter 7).[42-44] While an initial trial of medical therapy is reasonable and is often effective, earlier surgery may be beneficial in individuals with refractory seizures.[45,46] Earlier surgery offers no greater likelihood of successful seizure control than later surgery, but the cognitive outcome may be better with earlier surgery.[43] Even if both hemispheres are affected, resection of a severely affected region of one hemisphere is sometimes beneficial.[47] When the brain lesion is extensive, corpus callosotomy may reduce the number of atonic or secondarily generalized seizures.[48] Some individuals who are not candidates for lesion resection may benefit from a vagus nerve stimulator.

Hemiparesis often develops acutely in conjunction with the first flurry of seizures, and it may be difficult to determine whether such a focal deficit represents postictal paralysis or ischemic dysfunction.[49] This initial hemiparesis may or may not be transient. In other individuals, hemiparesis develops in a saltatory fashion over time, sometimes in the absence of seizures.[50] Children with early-onset unilateral weakness often exhibit impaired growth of the affected extremities, resulting in asymmetry of the two sides.

Infants with Sturge-Weber syndrome typically develop normally for several months after birth, although neurological impairment can be obvious even at birth, especially in babies with extensive bilateral brain involvement. Mental deficiency eventually occurs in about half of these individuals, and only 8% of those with bilateral brain involvement are intellectually normal.[11,21,51] The extent of mental impairment varies, however, and even patients with normal intelligence sometimes develop social, psychological, or behavioral problems (see Chapter 5).[52]

Intracranial hemorrhage is rare in individuals with Sturge-Weber syndrome. In 1906, Cushing described three patients whose acute onset seizures and hemiparesis he attributed to the hemorrhage from the syndrome without convincing evidence.[53] One adult woman developed subarachnoid hemorrhage, but it was not clear whether her hemorrhage was due to Sturge-Weber syndrome.[54] A few other reports of hemorrhage among patients with Sturge-Weber syndrome have appeared, but most of these cases were poorly documented, with insufficient pathological evidence to confirm hemorrhage or with unconvincing clinical evidence.

Aside from hemiparesis, various other focal neurological deficits may occur. The most common is a visual field defect due to the involvement of one or both occipital lobes or optic tracts with the leptomeningeal angiomatosis. Hydrocephalus may occur as the result of increased venous pressure from thrombosis of the deep venous channels or extensive arteriovenous anastomoses.[21,55,56] The neurological deficits in patients with Sturge-Weber syndrome eventually stabilize, leaving varying degrees of residual hemiparesis, hemianopia, cognitive or behavioral impairment, or epileptic seizures.[57] Individuals with no neurological dysfunction aside from epilepsy are relatively common.

Some patients undergo a series of step-wise, stroke-like events without seizures or intercurrent illness.[50] The mechanisms for this deteriorating neurologic function are still debated.[49] Potential factors include some form of hypoxic or ischemic injury to parenchymal brain tissue adjacent to the leptomeningeal angioma, venous occlusion, and the effects of increased venous pressure on the brain parenchyma.

DIAGNOSIS OF STURGE-WEBER SYNDROME

The diagnosis of Sturge-Weber syndrome depends on the facial and intracranial angiomas. Although the facial port-wine stain is the most obvious of the possible manifestations of Sturge-Weber syndrome, the leptomeningeal angioma is clearly the most important component in determining the ultimate prognosis of the patient. A few patients have the leptomeningeal angioma but not a facial angioma. These patients may display any of the usual neurologic manifestations of Sturge-Weber syndrome, and some of them have glaucoma or other ocular features. There seems to be no good reason to exclude these patients from the larger group of Sturge-Weber syndrome patients as they share the major clinical (neurological) manifestations of the syndrome. Usually, the patients without facial lesions are not identified until or unless they develop epilepsy or other neurologic problems.

At the other end of the spectrum, most of the children with facial port-wine stains do not have intracranial leptomeningeal angiomas.[15] Before the onset of neurologic symptoms and signs, the best way to document an intracranial leptomeningeal angioma is with one or more of the available neuroimaging studies (see Chapter 6).[10] Use of plain skull films has been supplanted by computed cranial tomography (CT) and magnetic resonance imaging (MRI). Computed tomography reliably depicts brain calcification, but MRI yields more precise pictures of the brain structure. Magnetic resonance imaging with gadolinium usually shows both the presence and the extent of the intracranial angioma, although the findings can be subtle at times.[58] Negative findings, especially in the neonatal period, do not rule out the presence of an intracranial angioma because the lesion may not be evident at that time. After a lesion is identified, delineation of its extent may be of value both for diagnosis and in consideration of possible epilepsy surgery.[59,60]

TREATMENT OF STURGE-WEBER SYNDROME

Improved treatment methods for many of the complications of Sturge-Weber syndrome have been developed over the last several years, making it possible to alter some aspects of the natural history of the syndrome. This possibility and the lack of any systematic compilation of the potential treatments are two of the reasons this volume was undertaken. Details of the therapies available for the cutaneous and ophthalmological complications of the disease are found in Chapters 2 and 3.

Medical management of the neurological complications is often difficult and sometimes impossible (see Chapters 4 and 5). Anticonvulsant medications may effectively limit the number of seizures, but these agents are not uniformly effective, especially in individuals with Sturge-Weber syndrome. The role of antiplatelet agents is uncertain and needs additional study. Individuals with medically refractory epilepsy often benefit from lesion resection (see Chapter 7), but surgery may not be feasible when the lesion is extensive.

Fundamental to the management of Sturge-Weber syndrome patients are unanswered questions about the mechanisms of the progressive loss of motor function seen in many of the children with this condition. Does this loss represent the cumulative effects of repeated thrombotic events within the leptomeningeal angioma, or is it the result of chronically disturbed blood flow dynamics and oxygen delivery to the tissues involved? Sturge-Weber syndrome is suspected to result from a somatic mutation of one or more genes that control vascular development, but if this is correct, which gene is affected? Answers to these types of questions may be prerequisite to the development of effective therapy. In this book we hope to compile recent information from various disciplines involved in the diagnosis and management of Sturge-Weber syndrome. Perhaps this effort will arouse additional interest among investigators who may begin to answer some of the fundamental questions surrounding this enigmatic condition.

REFERENCES

1. Schirmer R. Ein fall von telangiektasia. *Graefes Arch Ophthalmol* 1860;7:119-121.

2. Jennings Milles W. Naevus of the right temporal and orbital region: naevus of the choriod and detachment of the retina in the right eye. *Trans Ophthalmol Soc UK* 1884;4:168-171.

3. Sturge WA. A case of partial epilepsy, apparently due to a lesion of one of the vaso-motor centres of the brain. *Trans Clin Soc London* 1879;12:162-167.

4. Kalischer S. Demonstration des gehirns eines kindes mit telangiectasie der linksseitigen gesichts-kopfhaut und hirnoberfläche. *Berl Klin Wochenschr* 1897;34:1059.

5. Weber FP. Right-sided hemihypotrophy resulting from right-sided congenital spastic hemiplegia, with a morbid condition of the left side of the brain revealed by radiograms. *J Neurol Psychopath* 1922;3:134-139.

6. Dimitri V. Tumor cerebral congenito (angioma cavernoso). *Rev Assoc Med Argentina* 1923;36:1029-1037.

7. Krabbe KH. Facial and meningeal angiomatosis associated with calcification of the brain cortex. *Arch Neurol Psychiat* 1934;32:737-755.

8. Alexander GL, Norman RM. *Sturge-Weber Syndrome*. Bristol: John Wright & Sons Ltd, 1960.

9. Uram M, Zubillaga C. The cutaneous manifestations of Sturge-Weber syndrome. *J Clin Neuro Ophthalmol* 1982;2:245-248.

10. Roach ES. Neurocutaneous syndromes. *Pediatr Clin North Am* 1992;39:591-620.

11. Peterman AF, Hayles AB, Dockerty MB, Love JG. Encephalotrigeminal angiomatosis (Sturge-Weber disease). *J Am Med Assoc* 1958;167:2169-2176.

12. Kitahara T, Maki Y. A case of Sturge-Weber disease with epilepsy and intracranial calcification at the neonatal period. *Eur Neurol* 1978;17:8-12.

13. Rudolph J, Joubert M. Encephalo-angiomatosis in black children. A report of 2 cases. *S African Med J* 1984;65:93-97.

14. Hennedige AA, Quaba AA, Al-Nakib K. Sturge-Weber syndrome and dermatomal facial port-wine stains: incidence, association with glaucoma,

and pulsed tunable dye laser treatment effectiveness. *Plast Reconstr Surg* 2008;121:1173-1180.

15. Enjolras O, Riche MC, Merland JJ. Facial port-wine stains and Sturge-Weber syndrome. *Pediatrics* 1985;76:48-51.

16. Tallman B, Tan OT, Morelli JG et al. Location of port-wine stains and the likelihood of ophthalmic and/or central nervous system complications. *Pediatrics* 1991;87:323-327.

17. Ambrosetto P, Ambrosetto G, Michelucci R, Bacci A. Sturge-Weber syndrome without port-wine facial nevus - report of 2 cases studied by CT. *Childs Brain* 1983;10:387-392.

18. Crosley CJ, Binet EF. Sturge-Weber Syndrome- presentation as a focal seizure disorder without nevus flammeus. *Clin Pediatr* 1978;17:606-609.

19. Taly AB, Nagaraja D, Das S, Shankar SK, Pratibha NG. Sturge-Weber-Dimitri disease without facial nevus. *Neurology* 1987;37:1063-1064.

20. Boltshauser E, Wilson J, Hoare RD. Sturge-Weber syndrome with bilateral intracranial calcification. *J Neurol Neurosurg Psychiatry* 1976;39:429-435.

21. Bebin EM, Gomez MR. Prognosis in Sturge-Weber disease: comparison of unihemispheric and bihemispheric involvement. *J Child Neurol* 1988;3:181-184.

22. Roach ES, Riela AR. *Pediatric Cerebrovascular Disorders*. 2 ed. New York: Futura, 1995.

23. Royle HE, Lapp R, Ferrara ED. The Sturge-Weber syndrome. *Oral Surg Oral Med Oral Pathol* 1966;22:490-497.

24. Weiss DI. Dual origin of glaucoma in encephalotrigeminal haemangiomatosis. *Trans Ophthalmol Soc U K* 1973;93:477-493.

25. Phelps CD. The pathogenesis of glaucoma in Sturge-Weber syndrome. *Ophthalmology* 1978;85:276-286.

26. Christensen GR, Records RE. Glaucoma and expulsive hemorrhage mechanisms in the Sturge-Weber syndrome. *Ophthalmology* 1979;86:1360-1364.

27. Susac JO, Smith JL, Scelfo RJ. The "tomatoe-catsup" fundus in Sturge-Weber syndrome. *Arch Ophthalmol* 1974;92:69-70.

28. Cibis GW, Tripathi RC, Tripathi BJ. Glaucoma in Sturge-Weber syndrome. *Ophthalmology* 1984;91:1061-1071.

29. Sujansky E, Conradi S. Sturge-Weber syndrome: age of onset of seizures and glaucoma and the prognosis for affected children. *J Child Neurol* 1995;10:49-58.

30. Rosen E. Hemangioma of the choroid. *Ophthalmologica* 1950;120:127-149.

31. Witschel H, Font RL. Hemangioma of the choroid. A clinicopathologic study of 71 cases and a review of the literature. *Surv Ophthalmol* 1976;20:415-431.

32. Van Emelen C, Goethals M, Dralands L, Casteels I. Treatment of glaucoma in children with Sturge-Weber syndrome. *J Pediatr Ophthalmol Strabismus* 2000;37:29-34.

33. Eibschitz-Tsimhoni M, Lichter PR, Del Monte MA et al. Assessing the need for posterior sclerotomy at the time of filtering surgery in patients with Sturge-Weber syndrome. *Ophthalmology* 2003;110:1361-1363.

34. Olsen KE, Huang AS, Wright MM. The efficacy of goniotomy/trabeculotomy in early-onset glaucoma associated with the Sturge-Weber syndrome. *J AAPOS* 1998;2:365-368.

35. Budenz DL, Sakamoto D, Eliezer R, Varma R, Heuer DK. Two-staged Baerveldt glaucoma implant for childhood glaucoma associated with Sturge-Weber syndrome. *Ophthalmology* 2000;107:2105-2110.

36. Chevrie JJ, Specola N, Aicardi J. Secondary bilateral synchrony in unilateral pial angiomatosis: Successful surgical management. *J Neurol Neurosurg Psychiatry* 1988;15:95-98.

37. Barbagallo M, Ruggieri M, Incorpora G et al. Infantile spasms in the setting of Sturge-Weber syndrome. *Childs Nerv Syst* 2009;25:111-118.

38. Miyama S, Goto T. Leptomeningeal angiomatosis with infantile spasms. *Pediatr Neurol* 2004;31:353-356.

39. Oakes WJ. The natural history of patients with the Sturge-Weber syndrome. *Pediatr Neurosurg* 1992;18:287-290.

40. Kossoff EH, Ferenc L, Comi AM. An infantile-onset, severe, yet sporadic seizure pattern is common in Sturge-Weber syndrome. *Epilepsia* 2009;50:2154-2157.

41. Kramer U, Kahana E, Shorer Z, Ben-Zeev B. Outcome of infants with unilateral Sturge-Weber syndrome and early onset seizures. *Dev Med Child Neurol* 2000;42:756-759.

42. Rosen I, Salford L, Starck L. Sturge-Weber disease-neurophysiological evaluation of a case with secondary epileptogenesis, successfully treated with lobe-ectomy. *Neuropediatrics* 1984;15:95-98.

43. Bourgeois M, Crimmins DW, de Oliveira RS et al. Surgical treatment of epilepsy in Sturge-Weber syndrome in children. *J Neurosurg* 2007;106:20-28.

44. Arzimanoglou AA, Andermann F, Aicardi J et al. Sturge-Weber

syndrome: indications and results of surgery in 20 patients. *Neurology* 2000;55:1472-1479.

45. Roach ES, Riela AR, Chugani HT, Shinnar S, Bodensteiner JB, Freeman J. Sturge-Weber syndrome: recommendations for surgery. *J Child Neurol* 1994;9:190-193.

46. Falconer MA, Rushworth RG. Treatment of encephalotrigeminal angiomatosis (Sturge-Weber disease) by hemispherectomy. *Arch Dis Child* 1960;35:433-447.

47. Hallbook T, Ruggieri P, Adina C et al. Contralateral MRI abnormalities in candidates for hemispherectomy for refractory epilepsy. *Epilepsia* 2010;51:556-563.

48. Rappaport ZH. Corpus callosum section in the treatment of intractable seizures in the Sturge-Weber syndrome. *Child Nerv Sys* 1988;4:231-232.

49. Jansen FE, van der Worp HB, van HA, van Nieuwenhuizen O. Sturge-Weber syndrome and paroxysmal hemiparesis: epilepsy or ischaemia? *Dev Med Child Neurol* 2004;46:783-786.

50. Garcia JC, Roach ES, McLean WT. Recurrent thrombotic deterioration in the Sturge-Weber syndrome. *Childs Brain* 1981;8:427-433.

51. Aicardi J, Arzimanoglou A. Sturge-Weber syndrome. *International Pediatrics* 1991;6:129-134.

52. Chapieski L, Friedman A, Lachar D. Psychological functioning in children and adolescents with Sturge-Weber syndrome. *J Child Neurol* 2000;15:660-665.

53. Cushing H. Cases of spontaneous intracranial hemorrhage associated with trigeminal nevi. *JAMA* 1906;47:178-183.

54. Anderson FH, Duncan GW. Sturge-Weber disease with subarachnoid hemorrhage. *Stroke* 1974;5:509-511.

55. Fishman MA, Baram TZ. Megalencephaly due to impaired cerebral venous return in a Sturge-Weber variant syndrome. *J Child Neurol* 1986; 1:115-118.

56. Orr LS, Osher RH, Savino PJ. The syndrome of facial nevi, anomalous venous return and hydrocephalus. *Ann Neurol* 1978;3:316-318.

57. Pascual-Castroviejo I, Pascual-Pascual SI, Velazquez-Fragua R, Viano J. Sturge-Weber syndrome: study of 55 patients. *Can J Neurol Sci* 2008; 35:301-307.

58. Juhász C, Chugani HT. An almost missed leptomeningeal angioma in Sturge-Weber syndrome. *Neurology* 2007;68:243.

59. Chugani HT, Mazziotta JC, Phelps ME. Sturge-Weber syndrome: a study of cerebral glucose utilization with positron emission tomography. *J Pediatr* 1989;114:244-253.

60. Chiron C, Raynaud C, Tzourio N et al. Regional cerebral blood flow by SPECT imaging in Sturge-Weber disease: an aid for diagnosis. *J Neurol Neurosurg Psychiatry* 1989;52:1402-1409.

Chapter 2

Port Wine Stains and the Sturge-Weber Syndrome

Joanna M. Burch, M.D.

Joseph G. Morelli, M.D.

RISK OF THE STURGE-WEBER SYNDROME

Although cutaneous facial port-wine stains are a hallmark of Sturge-Weber syndrome, most children with a facial port-wine stain do not have the full syndrome. It is true that most individuals with Sturge-Weber syndrome have a port-wine stain in the area innervated by the first (ophthalmic) division of the trigeminal nerve, although the dermatomal location of the nevus may be coincidental.

It is important to know which facial port-wine stains are most likely to be associated with the syndrome. For a patient with any facial port wine stain, the overall risk of having Sturge-Weber syndrome is only 3% to 9%.[1-4] The likelihood of having Sturge-Weber syndrome increases to about 15% to 20% when the upper half of the face is involved and increases further in individuals whose nevus affects the upper face bilaterally. Although the majority of individuals with the Sturge-Weber syndrome have a cutaneous facial port-wine stain, a few people have its neurologic or ophthalmologic manifestations but lack a port-wine stain.

TREATMENT OF PORT WINE STAINS WITH THE VASCULAR SPECIFC PULSED DYE LASER
Historical Perspective

The treatment of port wine stains is based on the concept of selective photothermolysis elucidated in the mid-1980's by Anderson and Parrish.[5] This led to the development of the vascular specific pulsed dye laser targeting the beta absorption peak of hemoglobin at 577 nm. This type of laser has been commercially available and used for the treatment of port wine stains for over 20 years.

Initial studies in children and adults resulted in complete clearing with an average of 6.5 treatments.[6,7] Children less than seven years old at the beginning of therapy required slightly fewer treatments to achieve total clearing.[7] The laser pulse duration used in these studies was 300 and 360 microseconds respectively. With time it was noted that not all patients achieved complete clearing using these parameters.[8,9] A 360 or longer microsecond pulse duration is superior to shorter pulse durations.[8] Lesions on the trunk and extremities respond slightly better to treatment than facial lesions.[9]

One of the reasons for lack of complete clearing in all patients was the minimal penetration of 577 nm light into the dermis. Lack of clearing may be secondary to deeper vessels in the port wine stain being beyond the reach of the 577 nm light. It was demonstrated that in normal albino pig skin, light of 585 nm penetrated to a depth of 1.2 mm into the dermis, while light of 577 nm penetrated only 0.5 mm.[10] This same study showed that 590 nm light penetrated only to a depth of 0.8 mm and was less selective in vascular destruction. Histologic examination after treatment of adult port wine stain patients using both 577 nm and 585 nm revealed that the latter penetrated deeper in actual lesions.[11] Following these findings new lasers were created that emitted only 585 nm light, and the 577 nm units became obsolete. Also, the pulse duration of the laser was increased from 360 microseconds to 450 microseconds to correspond more closely with the predicted time of heating necessary to destroy the average vessel size in the port wine stains.

With a 585 nm, 450 microsecond laser, the overall lightening of port wine stains after one treatment in children aged 3 months to 14 years was 53% in one report.[12] Subsequent treatments generated diminishing results. Children 6 years of age or less responded slightly better than older children, but only 3 of 33 patients achieved total clearance. The location and size of lesions with total clearing was not provided, although these lesions were described as light pink. Lesions in the periorbital region and the neck improved more than lesions in other locations. Ashinoff and Geronemus reported that the use of this laser was safe in children form 6 weeks to 30 weeks of age.[13] Their results were similar to those of the previous study, but none of the port wine stains were cleared after an average of 2.8 treatments. In a series of 49 port wine stains in 43 children aged 2 weeks to 14 years, 16% had greater than 95% clearing and the average improvement was 69%.[14] Children less than 4 years of age responded better to therapy than older children, and lesions on the face, neck and torso responder better than extremity lesions.

Morelli and colleagues were the first to report that the size of the port wine stain predicted the response to treatment.[15] They demonstrated a 22% total clearing rate in 85 children. However, 32% of the port wine stains that were initially smaller than 20 cm^2 were cleared completely, while only 8% of patients with larger part wine stains had complete clearing. They also noted that children younger than 1 year of age responded better to therapy than older children (18% versus 32% with total clearing).

They expanded this work by comparing the response of facial port wine stains by location on the face, size of the lesions, and age of the patient at the time treatment was initiated.[16] Rather than use a subjective measure of lightening, they used objective decrease in size as a measurement of improvement. They determined that port wine stain location on the central forehead was the most important predictor of treatment response but that smaller lesion size and younger age at the beginning of treatment were also important. They also noted diminishing treatment effectiveness, with markedly greater improvement from the initial five treatments than the subsequent five treatments. **(Table 2-1 and Figure 2-1)**

Katugampola and Lanigan described a mixed pediatric and adult cohort of 640 patients from the United Kingdom.[17] In their analysis, facial port wine stain responded much better to treatment than non-facial lesions (58% versus 18% of the patients achieving 75% overall lightening). In contrast to other studies, 64% of their patients over 50 years of age received an "excellent" result compared to only 19% of those less than 5 years. In this series, the initial lesion color held no prognostic value. Similarly, another study of the effect of treatment on 89 patients with head and neck port wine stains divided the patients into four groups (0 to 5 years, 6 to 11 years, 12 to 17 years, and 18 to 31 years). There was a 40% reduction in color in all groups after an average of 5 treatments.[18]

These studies are difficult to compare because they do not use consistent parameters. The ages compared in the studies vary. Also, most of the studies report on qualitative lightening of the port wine stain and not on decrease in size of the lesion. Most reports do not spell out the initial size of the lesions in their comparisons or the exact location of the lesions, although it had been shown that these were very important factors in treatment success. Despite these study limitations, it became apparent that very young children could be safely treated; smaller lesions were more likely to be totally obliterated; lesions in the central forehead responded better than lesions in other areas; the initial

Table 2-1. Response of Port-Wine Stains to the First Five and Second Five Pulsed Dye Laser Treatments According to the Location of the Port-Wine Stain, Its Size, and the Age of the Patient at the Beginning of Treatment.*

	Patients (n)	Mean decrease in size of stain	
		First 5 Rx	Second 5 Rx
All patients	91	55%	18%
Location			
Forehead	13	100%	N/A
Lateral	39	58%	28%
Central	27	48%	14%
Mixed	12	21%	9%
Size			
<20 cm2	49	67%	21%
20-40 cm2	29	45%	8%
>40 cm2	13	23%	29%
Age			
<1 yr	30	63%	33%
1-6 yr	33	48%	15%
>6 yr	28	54%	10%

*Adapted from the *British Journal of Dermatology*.[16]

treatments achieved the most lightening of lesions with subsequent diminishing returns; prediction of treatment outcomes in any individual was difficult; and only a small percentage of lesions would clear totally.

The observation that only a finite improvement was attainable in most lesions spurred additional research. Superficial vessels of 150 microns and smaller were selectively destroyed by treatment, while deeper larger vessels were unaffected.[19,20] It had been suggested that the optimal pulse for vessel destruction would be between 1 and 10 microseconds,[21] and it was postulated that longer pulse durations and wavelengths greater than 585 nm might improve clearing of port wine stains.

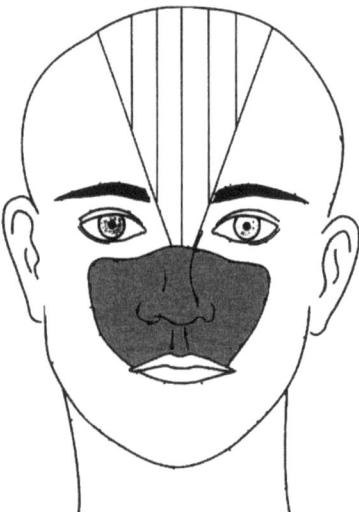

Figure 2-1. The solid area represents the central facial region; the striped area represents the central forehead region; all other areas are considered peripheral. *Reprinted with permission from Nguyen et al.[16]*

Comparison Studies

Few studies compare different wavelengths and pulse durations, and none of the comparison studies use similar patient selection criteria or laser settings. An early study compared 585 nm to 600 nm wavelength on test areas in 22 patients with port wine stains.[22] They concluded that 585 nm was the superior wavelength. Scherer and colleagues compared 585 nm with 590 nm, 595 nm, and 600 nm, and also 585 nm with 450 microsecond and 1.5 millisecond pulses.[23] Thirteen patients had the greatest improvement with 585 nm, while only 3 had maximal improvement with 590 nm, 8 with 595 nm and 6 with 600 nm. In the same study, a comparison of test sites in 62 patients demonstrated equality between 585 nm using either 450 microsecond or 1.5 millisecond pulse durations. Another study evaluated 2 treatments with 595 nm, 1.5 milliseconds in 9 patients previously treated with 585 nm, 450 microseconds.[24] Three of these patients showed a good response, 3 had a fair response, and 3 had little or no response. Another group compared 585 nm to 595 nm both with a 1.5 millisecond pulse duration in 1 to 6 treatments in 62 Asian patients.[25] These authors concluded that 585 nm was markedly superior to 595 nm.

Other investigators concluded that 585 nm, 500 microseconds was superior to either 595 nm, 500 microseconds or 595 nm, 20 milliseconds.[26] Another study retrospectively compared 585 nm, 1.5 millisecond with 595 nm with 1.5, 6, and 20 milliseconds in 18 patients averaging 35 years of age.[27] They concluded that 585 nm and 595 nm wavelengths with a 1.5 millisecond pulse were equivalent and that 595 nm with the longer pulse durations offered no clear advantage. Taken together, these studies suggest that, despite theoretical predictions, a 585 nm wavelength with either a 450 microsecond or 1.5 millisecond pulse duration is the best option.

A study of 21 patients with an incomplete response to 450 microsecond therapy at 585 nm showed further improvement after treatment with a 585 nm, 1.5 millisecond laser.[28] The same investigator next evaluated 20 patients who had become "refractory" to 585 nm, 450 microsecond treatment using 595 nm, 1.5 millisecond settings.[29] He concluded that with 595 nm, 1.5 millisecond, a 76% improvement was obtained in an average of 3.1 treatments compared to only a 40% improvement with an average of 8.8 previous treatments using 585 nm, 450 microseconds.

A Japanese group evaluated 40 patients who had become "resistant" to 585 nm, 450 microsecond therapy using a 595 nm, 1.5 millisecond laser.[30] Eight of their patients showed excellent improvement, 9 had good improvement, 11 had fair improvement, and 12 had poor improvement. None of these studies mentioned the size or exact location of the lesions, so despite some effort to evaluate optimal laser parameters it is still difficult to make the best choice. Both 585 nm and 595 nm and 450 microseconds and 1.5 millisecond pulse durations are effective. At this time there are no pretreatment methods of evaluation which allow for optimal parameter selection. Still, the lack of complete clearing in the majority of patients remains a problem.

Newer Techniques

Because most port wine lesions are not completely obliterated with the commonly used techniques, two new methods are being investigated. Hybrid lasers combine a pulsed dye laser at 595 nm with a neodymium: yttrium aluminum garnet (Nd:YAG) laser at 1064 nm. The 595 nm laser is triggered first and followed in one second by a pulse from the Nd:YAG. The novel aspect of this approach is the combination of two lasers into one unit that controls the time of the administration of each laser. So far there is little experience with this technique.[31, 32]

Port wine lesions are vascular malformations and not tumors, but they are not static. Untreated, they gradually become a darker color, thicken, and develop superficial nodules and blebs. Even with treatment, this tendency to progress remains, sometimes resulting in marked worsening of previously improved lesions. The use of topical angiogenesis inhibitors might decrease the likelihood of lesion progression and lessen the chances of revascularization after treatment.[33, 34] Systemic rapamycin decreases facial angiofibromas in individuals with tuberous sclerosis complex, perhaps providing a precedent for drug treatment or port wine lesions.[35] To make this approach plausible for use in port wine stains will require the development of effective topical angiogenesis inhibitors.

Cooling

Oxygenated hemoglobin in the ectatic dermal capillaries of a port wine stain is the target chromophore for the pulsed dye laser, but epidermal melanin also absorbs light at this wavelength. Energy absorption by melanin leads to heating of the epidermis and increases the likelihood of crusting, blistering, and scarring. Absorption of the more superficial melanin also reduces the laser light energy reaching the deeper vessels targeted for treatment. Earlier attempts to cool the epidermis during laser therapy (e.g. with ice cubes or chilled water) seemed to prevent pain but did not improve treatment efficacy. Prolonged cooling with these methods produces a steady state cooling effect which chills the abnormal dermal vessels as well as the epidermis, and any protective effect on the epidermis may be offset by the additional energy required to achieve the same clinical result on the now-cooled blood vessels.[36]

Several authors treated PWS with a dynamic cooling device (DCD) that applied a localized cryogen spurt (typically dichloro-difluoro-methane or 1,1,1,2-tetrafluoro-ethane) to the epidermis for a very short (5-80 millisecond) time in order to selectively cool the epidermis while leaving the temperature of the underlying PWS vessels unchanged.[36-41] Looking at skin temperature measurements using a fast infrared imaging detector, Nelson and associates showed that the surface temperature of the skin prior to laser exposure could be decreased by as much at 40 degrees C using the DCD.[36] They chose 18 test sites in representative areas of PWS in 14 subjects and treated all with the PDL, 585nm, pulse duration 0.45 milliseconds (ms), 5mm spot size, at the maximum possible light dosage of 10 J/cm^2. Cryogen spray was applied to a 7mm spot area corresponding to the laser test sites for 5-80 ms spurts and compared to uncooled sites. The comparative skin surface temperature measurements showed that in skin cooled prior to the laser treatment, the maximum temperature reached immediately after the laser pulse is decreased and the return to normal temperatures is more rapid. Epidermal necrosis was evident in 57% of the areas treated with the five-ms cryogen spurt duration, and 18% of the areas treated with the 10-ms duration. The authors concluded that dynamic epidermal cooling using a 20-80-ms cryogen spurt permits exposure of PWS skin to a dosage (10 J/cm^2) that was proven to cause epidermal necrosis in uncooled treated sites.[36] Patients also reported much less pain with the cryogen cooling; patients who had durations longer than 30-40 milliseconds often reported feeling no pain at all. Six months after laser exposure, "clinically significant" blanching of the PWS on all cooled sites was noted. The authors reasoned that this indicates that the epidermis was cooled selectively without affecting the temperature of the deeper PWS vessels.

Waldorf and colleagues looked at 47 patients undergoing PDL treatment for PWS and the effect of cryogen spray cooling (CSC) on pain, clearance of PWS, and pigment changes.[37] One site within the PWS was treated without CSC and another site with a 40-millisecond spurt of cryogen just before the laser pulse. Both sites chosen were thought to be "representative" of the PWS. The fluences applied with the CSC were increased by 10% to 20%. Pretreatment spectrometer readings were taken and repeated at each visit (3 visits 6-8 weeks apart). Photographs were also taken for comparison of clinical clearance of the PWS. There was not a statistically significant difference in the percent change from baseline of the spectrometer readings between the group with or without

CSC. Clinically there was equal clearing of the treated areas of PWS in the groups based on evaluation of the photographs. Post treatment purpura was similar for the "majority of the patients" in both groups. Four patients had textural or pigment changes and these adverse affects happened in both groups and not in any particular skin type. Nor were the side effects related to higher fluences during treatment. The authors related this to possible spurt-to-spurt variation of cryogen spray amounts caused by air bubbles in the hose near the nozzle of the DCD. There was a statistically significant reduction in pain reported by patients with CSC, most pronounced in darker skin types (III and IV). When the authors looked at pain reduction by age (>12 years versus <12 years), the effect on pain reduction by the CSC was less for younger children, possibly due to an increased level of anticipatory anxiety in this age group combined with the scary "hissing" sound of the cryogen spurt. The authors concluded that the CSC decreased pain and did not compromise efficacy. Whether the CSC had a protective effect on the epidermis was unable to be determined from this study.

In a retrospective study of 196 Asian patients, Chang and Nelson[38] analyzed differences in measured blanching response scores in 98 patients treated without CSC at 5-7 J/cm^2 before June 1995 versus 98 patients treated with CSC at fluences of 8-10 J/cm^2 after June 1995. The lesions in both groups were stratified by severity (1-faint, barely discernible borders with areas of normal skin interspersed within the lesion and 2-well defined borders, uniform lesion with no areas of normal skin, plus raised or thickened lesion with modularity or hypertrophy of anatomic structure).

The higher fluences with CSC did not result in increased clearing in lesions of severity score 1. In the patients with severity score 2 PWS, there was significantly more clearing in the higher fluence with CSC group. Reasons for this finding were not discussed. Hyperpigmentation occurred in 57% of patients without and 48% of the patients with CSC. This resolved spontaneously within one year in all patients. Two patients treated without CSC developed permanent hyperpigmentation. Two patients with CSC had transient, self-resolving hypertrophic scarring, while 3 (3.1%) of the patients treated without CSC developed permanent hypertrophic scarring.

Pain was not assessed in this study. Because the side effects were similar and the higher fluences with CSC were more effective overall, the authors recommend using the higher fluences with CSC. Nelson, and colleagues reported a 6 month follow up of 14 patients recruited from the same location as in his previous report.[36, 39] The authors reported clinically similar clearing in areas with and without CSC. Adverse effects, including hypertrophic scarring, changes in pigmentation, atrophy, or induration occurred on 10%, 25% and 60% of the uncooled sites exposed to light dosages of 8, 9, and 10 J/cm², respectively. The CSC-pretreated sites had no skin surface textural changes. There was "clinically equivalent, significant blanching on all cooled test sites," which implies adequate heating of the PWS vessels with laser treatment with CSC. This study suggests that there is an epidermal protective effect of the CSC at higher fluences (8-10 J/cm²) without compromising efficacy.

Using a DCD to deliver cryogen to the skin has the theoretic risk of misfiring in both timing and amount, resulting in damage to the epidermis at the higher laser fluences. Dichloro-difluoro-methane depletes the ozone layer,[40,42] although 1,1,1,2-tetrafluoro-ethane is a chlorine-free hydrofluorocarbon that is ozone friendly. The cryogen canisters must be replaced and are expensive. An alternative to address these concerns is to use a constant-flow cooling device (Zimmer Elektromedizin, Germany) which uses a compressor to take room air and produce a cold stream of air with a flow of 500 to 1,000 L/min and a temperature as low as −30 degrees C. Cold air cooling (CAC) incurs less operating costs and the risk of pulse-to-pulse failure or variation of cooling is eliminated. Many patients do not like the high airflow around the nose and mouth, although this can be improved by covering these areas and closing the nose during perinasal treatments. CAC involves a continuous and constant flow of cool air applied to the skin, which has the potential to create the steady-state cooling effect that could potentially cool the superficial vessels of the PWS in addition to the epidermis. Raulin and colleagues measured air and skin temperatures with an infrared thermometer using CAC, and found an air temperature of −15 degrees C at cooling level 6. Skin temperatures with the CAC were 28 degrees C after 1 second and 15 degrees C after 8 seconds (with an initial skin temperature of 32 °C).[41] The authors mention that skin temperatures are likely lower in reality because of the pretreatment cooling of the skin in adjacent areas, caused by the application of the CAC device while they are being treated.

CSC with an 80-ms spurt applied to a skin temperature of 30 degrees C leads to a decrease in skin temperature to -10 degrees C.[39] Although these measurements seem to indicate that CSC cools the skin to lower temperatures and for shorter periods, no comparative studies evaluating CSC versus CAC have been done. Several small studies have been done utilizing cold air cooling (CAC).[41-43] Raulin and associates analyzed analgesia produced by CAC in 166 patients with multiple diagnoses treated with multiple lasers (21 patients of those were treated with PDL for hemangiomas, telangiectasias, and PWS) and 87% of patients clearly preferred CAC versus ice gel.[41] The number increased to 96% if perinasal areas were excluded. They were able to increase energy treatment levels by 15% to 30% while short term side effects like purpura, crusting, and erythema were decreased in the CAC group. The efficacy of the treatment with the CAC was not addressed in this study.

Another prospective study summarized 13 PWS patients, comparing clearance, analgesia, and adverse effects utilizing a set PDL laser treatment (585nm, 0.45ms, 7mm spot size, 6J/cm2) with CAC (level 6) on 50% of the PWS and without cooling on the other 50%.[42] In larger stains, a 10 x 10cm treatment area was chosen and divided 50-50 for treatment in the study. For patients with CAC, clearance rates were similar to those without cooling in 9 (69%), better in 2 (15%), and worse in 2 (15%). Ninety-two percent of patients felt the analgesia the same or better than without cooling. Short-term adverse effects like purpura, erythema, and edema were less marked in the cooled sites. Dyspigmentation was less with cooling and no scarring occurred. The authors discuss possible reasons for the decreased clearing with cooling, given that the fluence was not increased. This tendency has been noted in other studies where there was not an increase in the fluence to accompany the cooling of the laser site.[39,42,43]

Hammes and colleagues address this issue in a study of 11 patients by comparing three laser settings applied within a single PWS in each patient.[43] The stains were divided into thirds, and were treated with 1: 6 J/cm² without CAC, 2): 6 J/cm² with CAC, and 3): 9 J/cm² with CAC. There was a slight but statistically insignificant increase in clearance in group 3 compared to group one. However, there was a decrease in clearance in group 2 compared to group 1, leaving the fluence the same but adding air decreased clearance in this small study. Side effects occurred at the same rate with the higher fluence as with the lower fluence with or without CAC. There was no scarring in any group. Pain was significantly decreased in both groups with cooling. When the authors looked at the color of the PWS treated (pink, red, or purple), red lesions showed the best clearance rates. The color and thickness of the stain influences response to PDL therapy.[38]

Most studies done with cooling in PWS have looked at the 585nm PDL. One study in infants showed efficacy in treating PWS utilizing a 595nm wavelength, 1.5ms pulse duration, energies of 4-8 J/cm², and a 40 ms duration spurt of cryogen 40ms prior to the laser pulse.[44]

The optimum laser and cooling settings are not well established by large studies. The studies that have been done seem to agree that analgesia is improved overall with cooling, and higher fluences are able to be used without increasing frequency of side effects utilizing cooling methods. Cooling clearly improves patient comfort. Caution should still be taken in treating patients with very dark stains, and those with darker skin types.[40]

Treatment Side Effects

Pulsed dye laser therapy for PWS is generally thought to have a low incidence of side effects. Many studies have addressed the issue of side effects with this treatment,[45-50] but there is no standardized way to document side effects, and follow-up lengths and intervals vary from study to study. Some studies evaluated side effects after a test treatment and then one full treatment,[45,46] while others looked at side effects over multiple treatments.[47-50] In general, hyperpigmentation is the most common side effect of PDL therapy, ranging from 1% to 27%,[45-50] with an early study reporting the incidence as high as 57%.[7] In most cases the hyperpigmentation resolved spontaneously over 6 months. In one study in which three patients had biopsies of post-laser hyperpigmentation[48], all three showed evidence of dermal melanophages, not hemosiderin (Masson-Fontana, and Pearls stains were used to confirm). Hypopigmentation had a lower incidence in the vast majority of studies, ranging from 1% to 3.4%.[45-51] Hypotrophic scarring occurred in 0.8% to 3%[45,49,51] of patients with some reports noting resolution in up to 30% of the cases over 6 to 12 months.[48,49] Fortunately, hypertrophic scarring was the least likely complication (0 to 1.1%) in several series.[46-60,] In those with long-term follow up, no hypertrophic scarring resolved with time. Short-term side effects like erythema, edema, crusting, and less often, blistering, were found in most studies.[46,48-51] The incidence of side effects was not related to body site,[45,46] season[45] or recent change in laser energy.[45,50]

One study of 701 patients in which test sites were done before treatment, the authors found that test site areas were not predictive of scarring.[49] Wareham and colleagues analyzed side effects in 265 PWS patients who received 650 treatments, and calculated the side effect rate per treatment.[50] Their results showed an overall incidence of side effects of 1.4% per treatment. The incident per treatment of specific side effects were as follows: blistering (0.3%), hyperpigmentation (0.6%), hypopigmentation (0.3%), scabbing (less than 0.2%). In this study the rate of side effects per treatment in children was lower than the overall average (0.6%). Other side effects have been reported rarely, like pyogenic granuloma,[46] Koebner phenomena with viligo,[53] and atrophie blanche-like scarring on the lower legs.[54]

Haedersdal and colleagues did a risk assessment for side effects with PDL on the inner arms of 14 patients of different Fitzpatrick skin types.[55] The inner arm was divided into three laser treatment areas. One area was pretreated with nicotinic acid to induce redness, a second was pretreated with placebo cream, and a third was an untreated area of skin. Baseline skin pigmentation and redness was quantified by skin reflectance measurements in all three sites, which were then treated with PDL. The skin reflectance measurements were repeated at 3 and 6 months. The results indicated that laser-induced pigmentary alterations were provoked at lower fluences than texture changes for individuals with similar degrees of pigment and redness. The risk of inducing a clinically visible pigmentary alteration or texture change increased with higher preoperative skin pigmentation and redness, and with the application of increasing laser doses. This supports the idea that darker lesions in darker-skinned patients may be the highest risk for side effects. The reflectance measurements showed objectively that dyspigmentation at 3 months improved at 6 months, as we have seen in most of the clinical studies.

Eppley and Sadove analyzed the possibility of significant hemoglobinemia that could be triggered in infants and children 5 years or younger undergoing PDL treatment of large PWS.[56] Fifteen subjects undergoing PDL treatment under general anesthesia of large (>3% body surface area) PWS had pre- and post-laser treatment haptoglobin assays and post operative urine hemosiderin levels checked. While there was evidence of a small level of hemoglobinemia in the blood haptoglobin assays, it did not exceed the renal threshold for hemoglobin in any patient. No patient had urine hemosiderin. The authors concluded that the intra-vascular hemolysis in this setting is unlikely to represent a serious risk to most children.

Several studies done with cooling have indicated a trend toward the utility of higher fluences with similar or fewer side effects.[36,38,42] Newer lasers with different setting variability may decrease the adverse effects as well. A questionnaire-based study of 51 PWS patients treated with the V-beam PDL (595nm, 1.5ms) using a cryogen spray cooling device indicated that early post-treatment symptoms were similar in incidence to a historical comparison group treated with the 585nm, 0.45ms PDL, but lasted several days fewer. Long-term side effects were not evaluated.[57]

Risk of Nevus Recurrence

Although pulsed-dye laser therapy for port-wine stains has been available for more than two decades, few follow-up studies have documented the lesion recurrence rate after treatment. Worsening over time is often characterized as a recurrence, but we believe that it results instead from the natural tendency of the remaining undestroyed portion of the nevus to darken. Much of the evidence to indicate recurrence is anecdotal or based on subjective data, like physician or patient questionnaires. One of the first series to analyze recurrence was that of Orten and colleagues, who reported on response and recurrence in 118 PWS in 108 patients treated with pulsed-dye laser.[58] Thirteen of 102 patients (12.7%) had some recurrence of their PWS. When divided into groups based on the amount of time elapsed since their last treatment, the recurrence rate of the PWS in the groups of patients 3 to 4 years after the last treatment approached 50%. The recurrence rate increased with the amount of time since the completion of laser therapy. However, patient numbers in those longer-term follow up groups were very small. Mørk and colleagues reported a recurrence rate of 11% after "several" years of treatment.[59] Michel and associates sent follow-up questionnaires to patients who had completed PDL treatment of a PWS.[60] One hundred forty seven patients responded, and then were contacted and examined for direct comparison with pre- and post-treatment photos. Overall, 16% experienced redarkening of their PWS. These subjects were stratified by age, and there were no redarkening of PWS in the group that was less than 10 years of age. This was significantly different from the recurrence rate of the group over 10 years (19%). Redarkening was also related to the rated color at the beginning of treatment (pink, red, or purple). The initial color of the lesions that redarkened was significantly darker than those that did not. Treatment parameters (spot size or energy density), patient gender, or PWS location did not seem to influence recurrence in this study. Unlike the Orten *et al.*[58] series, early recurrence rates were higher and did not increase with the time since the last laser treatment.

The largest study published to date to use objective chromometer measurements to assess the long-term efficacy of PDL treatment of PWS was done by Huikeshoven and associates.[61] This was a follow-up study from their 1998 report that evaluated the effectiveness of five laser treatments of PWS with the PDL by comparing pre- and post-treatment color measurements of the darkest part of the PSW in 89 patients.[62] Fifty one patients from the previous study met criteria and agreed to have follow-up color measurements done. The median length of follow up since the first five treatments was 9.5 years. Forty-five of 51 patients had further laser treatment after the initial study at the same institution. In that group, a median of seven extra treatments were done, and it had been a median of 5.8 years since the last treatment. They measured the difference in color between the stain and normal skin (denoted ΔE) at long-term follow up (median 12.4), and compared it to the baseline (median 15.2) and post-treatment ΔE values (median 8.9). A ΔE value of 1 is the least noticeable difference by a human observer. A higher ΔE indicates a greater color difference between the normal skin and the stain.) Although 45 patients underwent extra treatment sessions, all patients had darkening of their stains (increase in ΔE) during the follow up period, but the values were not as high as the baseline values (the color was not as dark as before treatment). The authors concluded that while re-darkening occurs at long-term follow up after PDL treatment of PWS, the stains are still significantly lighter than at baseline. Patients should be informed of this reality prior to beginning treatment.

Based on additional questionnaire data, 59% of patients were satisfied with their treatments, and when comparing patient evaluations of changes in the color of the PWS compared to the objective measurements, the authors found that patients tended to underestimate the color change taking place in their stains, and theorized that this is likely because of the slow nature of that change. This emphasizes the need use objective measurements to assess recurrence rather than patient or physician questionnaires.

The exact mechanisms for recurrence of PWS after treatment with the PDL have not been objectively evaluated. Continual vascular ectasia over time,[63] the limited diameter vessels and depth that the PDL can reach in a PWS,[64] leaving larger and deeper vessels largely untreated and capable of angiogenesis, and the role of decreased innervation of the abnormal vessels with dysregulation of blood flow[65] likely contribute. Advances in pulsed-dye laser technology, which include multiple wavelengths, varying pulse widths, and dynamic cooling, allowing for increased energy densities, may change the efficacy of laser treatment of PWS.[66] However, there are no controlled comparative studies to indicate this at the present time.

REFERENCES

1. Ch'ng S, Tan ST. Facial port-wine stain-clinical stratification and risks of neuor-ocular involvement. J Plast Reconstr Aesthet Surg. 2008, 61:889-93

2. Hennedige AA, Quaba AA, Al-Nakib K. Sturge-Weber syndrome and dermatomal facial port-wine stains: incidence, association with glaucoma, and pulsed tunable de laser treatment effectiveness. Plast Reconstr Surg. 2008, 12:1173-80

3. Tallman B, Tan OT, Morelli JG, et al. Location of port-wine stains and the likelihood of ophthalmic and/or central nervous system complications. Pediatrics. 1991, 87:323-7

4. Enjolras O, Riche MC, Merland JJ. Facial port-wine stains and Sturge-Weber syndrome. Pediatrics. 1985, 76:48-51

5. Anderson RR, Parrish JA. Selective photothermolysis: precise

microsurgery by selective absorption of pulsed irradiation. Science. 1983,220:524-7

6. Morelli JG, Tan OT, Garden J, et al. tunable dye laser (577 nm) treatment of port wine stains. Laser Surg Med. 1986, 6:94-9

7. Tan OT, Sherwood K, Gilchrest BA. Treatment of children with port wine stains using the flashlamp-pulsed tunable dye laser. N Engl J Med. 1989, 320:416-21

8. Garden JM, Polla LL, Tan OT. The treatment of port-wine stains by the pulsed dye laser. Analysis of pulse duration and long-term therapy. Arch Dermatol. 1988, 124:889-96

9. Glassberg E, Lask GP, Tan EM, Uitto J. The flashlamp-pumped 577-nm pulsed tunable dye laser: clinical efficacy and in vitro studies. J Dermatol Surg Oncol. 1988, 14:1200-8

10. Tan OT, Murray S, Kurban AK. Action spectrum of vascular specific injury using pulsed irradiation. J Invest Dermatol. 1989, 92:868-71

11. Tan OT, Morrison P, Kurban AK. 585 nm for the treatment of port-wine stains. Plast Reconstr Surg. 1990, 86:1112-7

12. Reyes BA, Geronemus R. Treatment of port-wine stains durng childhood with the flashlamp-pumped pulsed dye laser. J Am Acad Dermatol. 1990, 23:1142-8

13. Ashinoff R, Geronemus RG. Flashlamp-pumped pulsed dye laser for port-wine stains in infancy: earlier versus later treatment. J Am Acad Dermatol. 1991, 24:467-72

14. Goldman MP, Fitzpatrick RE, Ruiz-Esparza J. Treatment of port-wine stains (capillary malformations) with the flashlamp-pumped pulsed dye laser. J Pediatr. 1993, 122:71-7

15. Morelli JG, Weston WL, Huff JC, Yohn JJ. Initial lesion size as a predictive factor in determining the response of port-wine stains in children treated with the pulsed dye laser. Arch Pediatr Adol Med. 1995, 149:1142-4

16. Nguyen CM, Yohn JJ, Huff JC, et al. Facial port-wine stains in childhood: prediction of the rate of improvement as a function of the age of the patient, size an dlocation of the port wine stain and the number of treatments with the pulsed dye (585 nm) laser. Br J Dermatol. 1998, 138:821-5

17. Katugampolsa GA, Lanigan SW. Five years' experience of treating port wine stains with the flashlamp-pumped pulsed dye laser. Br J Dermatol. 1997, 137:750-4

18. Van der Horst CM, Koster PH, de Borgie CA, et al. Effect of the

timing of treatment of port-wine stains with the flashlamp-pumped pulsed dye laser. N Engl J Med. 1998, 338:1028-33

19. Hohenleutner U, Hilbert M. Wlotzke U, Lanthaler M. Epidermal damage and limited coagulation depth with the flashlamp-pumped pulsed dye laser: a histochemical study. J Invest Dermatol. 1995, 104:798-802

20. Fikerstrand EJ, Svaasand LO, Kopstad G, et al. Photothermally induced vessel-wall necrosis after pulsed dye laser treatment: lack of response in port-wine stains with small sized or deeply located vessels. J Invest Dermatol. 1996, 107:671-5

21. Dierickx CC, Casparian JM, Venugopian V, et al. thermal relaxation of port-wine stain vessels probed in vivio: the need for 1-10-millisecond laser pulse treatment. J Invest Dermatol. 1995, 195:709-14

22. Edstrom DW, Ros AM. The treatment of port-wine stains with the pulsed dye laser at 600 nm. Br J Dermatol 1997, 136:360-3

23. Scherer K, Lorenz S, Wimmershoff M, et al. Both the flashlamp-pumped dye laser and the long-pulsed tunable dye laser can improve results in port-wine stain therapy. Br J Dermatol. 2001, 145:79-84

24. Laube S, Taibjee S, Lanigan SW. Treatment of resistant port wine stains with the V-beam pulsed dye laser. Lasers Surg Med. 2003, 33:282-7

25. Chang CJ, Kelly KM, van Gemert MJ, Nelson JS. Comparing the effectiveness of 585-nm vs 595-nm wavelength pulsed dye laser treatment of port wine stains in conjunction with cryogen spray cooling. Lasers Surg Med. 2002, 31:352-8

26. Greve B, Raulin C. Prospective study of port wine stain treatment with dye laser: comparison of two wavelengths (585 nm vs. 595 nm) and two pulse durations (0.5 milliseconds vs. 20 milliseconds). Laers Surg Med. 2004, 34:168-73

27. Yung A, Sheehan-Dare R. A comparison study of a 595-nm with a 585-nm pulsed dye laser in refractory port wine stains. Br J Dermatol. 2005, 153:601-6

28. Bernstein EF, Brown DB. Efficacy of the 1.5 millisecond pulse-duration, 585 nm pulsed-dye laser for treating port-wine stains. Lasers Surg Med. 2005, 36:341-6

29. Bernstein EF. High energy 595 nm pulsed dye laser improves refractory port-wine stains. Dermatol Surg. 2006, 32:26-33

30. Kono T, Sakurai H, Takeuchi M, et al. Treatment of resistant port-wine stains with a variable-pulse pulsed dye laser. Dermatol Surg. 2007, 33:951-6

31. Borges da Costa J, Boixede P, Moreno C, Santiago J. Treatment of

resistant port-wine stains with a pulsed dual wavelength 595 and 1064 nm laser: a histochemical evaluation of the vessel wall destruction and selectivity. Photmoed Laser Surg. 2009, 27:599-605

32. Alster TS, Tanzi EL. Combined 595-nm and 1,064-nm laser irradiation of recalcitrant and hypertrophic port-wine stains in children and adults. Dermatol Surg. 2009, 35:914-8

33. Phung TL, Oble DA, Jia W, et al. Can the wound healing response of human skin be modulated after laser treatment and the effects of exposure extended? Implications on the combined use of the pulsed dye laser and a topical angiogenesis inhibitor for the treatment of port wine stain birthmarks. Lasers Surg Med. 2008, 40:1-5

34. Chang CJ, Hsiao YC, Mihm MC Jr, Nelson JS. Pilot study examining the combined use of pulsed dye laser and topical imiquimod versus laser alone for the treatment of port wine stain birthmarks. Lasers Surg Med. 2008, 40:605-10

35. Hofbauer GF, Marcollo-Pini A, Corsenc A, et al. The mTOR inhibitor rapamycin significantly improves facial angiofibroma lesions in a patient with tuberous sclerosis. Br J Dermatol. 2008, 159:473-5

36. Nelson JS, Milner TE, Anvari B, et al. Dynamic epidermal cooling during pulsed laser treatment of port-wine stain: A new methodology with preliminary clinical evaluation. *Arch Dermatol* 1995;131(6):695-700. PMID: 7778922.

37. Waldorf HA, Alster TS, McMillan K, et al. Effect of dynamic cooling on 585-nm pulsed dye laser treatment of port-wine stain birthmarks. *Dermatol Surg* 1997;23(8):657-62. PMID: 9256912.

38. Chang CJ, Nelson JS. Cryogen spray cooling and higher fluence pulsed dye laser treatment improve port-wine stain clearance while minimizing epidermal damage. *Dermatol Surg* 1999;25(10):767-72. PMID: 10594577.

39. Nelson JS, Milner TE, Anvari B, et al. Dynamic epidermal cooling in conjunction with laser-induced photothermolysis of port wine stain blood vessels. *Lasers Surg Med* 1996;19(2):224-9. PMID: 8887927.

40. White JM, Siegfried E, Boulden M, Padda G. Possible hazards of cryogen use with pulsed dye laser. A case report and summary. *Dermatol Surg* 1999;25(3):250-2. PMID:1019377.

41. Raulin C, Greve B, Hammes S. Cold air in laser therapy: First experiences with a new cooling system. *Lasers Surg Med* 2000;27:404-10. PMID: 11126434.

42. Greve B, Hammes S, Raulin C. The effect of cold air cooling on 585nm

pulsed dye laser treatment of port-wine stains. *Dermatol Surg* 2001;27(7):633-6. PMID: 11442613.

43. Hammes S, Roos S, Raulin C, Ockenfels HM, Greve B. Does dye laser treatment with higher fluences in combination with cold air cooling improve the results of port-wine stains? *J Eur Acad Dermatol Venereol* 2007;21(9):1229-33. PMID: 17894710.

44. Geronemus RG, Quintana AT, Lou WW, Kauvar AN. High-fluence modified pulsed dye laser photocoagulation with dynamic cooling of port-wine stains in infancy. *Arch Dermatol* 2000;136(7):942-3. PMID: 10891010.

45. Fiskerstrand EJ, Svaasand LO, Volden G. Pigmentary changes after pulsed dye laser treatment in 125 norther European patients with port wine stains. *Br J Deramtol* 1998;138(3):477-9. PMID:9580802.

46. Wlotzke U, Hohenleutner U, Abd-El-Raheem TA, Baumler W, Landthaler M. Side-effects and complications of flashlamp-pumped pulsed dye laser therapy of port-wine stains. A prospective study. *Br J Dermatol* 1996;134(3):145-30. PMID: 8731672.

47. Levine VJ, Geronemus RG. Adverse effects associated with the 577- and 585-nanometer pulsed dye laser in the treatment of cutaneous vascular lesions: A study of 500 patients. *J Am Acad Dermatol* 1885;32:613-7. PMID: 7896952.

48. Boixeda P, Nunez M, Perez B, et al. Complications of 585-nm pulsed dye laser therapy. *Int J Dermatol* 1997;36(5):393-7. PMID: 9199994.

49. Seukeran DC, Collins P, Sheehan-Dare RA. Adverse reactions following pulsed tunable dye laser treatment of port wine stains in 701 patients. *Br J Dermatol* 1997;136(5):725-9. PMID: 9205506.

50. Wareham WJ, Cole RP, Royston SL, Wright PA. Adverse effects reported in pulsed dye laser treatment for port wine stains. Lasers Med Sci 2009;24(2):241-6. PMID: 18418641.

51. Pence B, Aybey B, Ergenekon G. Outcomes of 532 nm frequency-doubled Nd:YAG laser use in the treatment of port-wine stains. *Dermatol Surg* 2005;31:509-517.

52. Gaston DA, Clark DP. Facial hypertrophic scarring from pulsed dye laser. Dermatol Surg 1998;24(5):523-5. PMID: 9598005.

53. Sommer S, Sheehan-Dare RA. The Koebner phenomenon in vitiligo following treatment of a port-wine stain naevus by pulsed dye laser. Br J Dermatol 1998;138(1):200-1. PMID: 9536256.

54. Sommer S, Sheehan-Dare RA. Atrophie blanche-like scarring after pulsed dye laser treatment. J Am Acad Dermatol 1999;41(1);100-2. PMID: 10411418.

55. Haedersdal M, Gniadecka M, Efsen J, et al. Side effects from the pulsed dye laser: the importance of skin pigmentation and skin redness. Acta Derm Venereol 1998;78(6):445-50. PMID: 9833046.

56. Eppley BL, Sadove AM. Systemic effects of photothermolysis of large port-wine stains in infants and children. Plast Reconstr Surg 1994;93(6):1150-3. PMID: 8171134.

57. Loffeld A, Zaki I, Abdullah A, Lanigan S. Study of patient-reported morbidity following V-beam pulsed-dye laser treatment of port wine stains. Lasers Med Sci 2005;20(3-4):114-6. PMID: 16047083.

58. Orten SS, Waner M, Flock S, Roberson PK, Kincannon J. Port-wine stains. An assessment of 5 years of treatment. *Arch Otolaryngol Head Neck Surg.* 1996;122(11):1174-9. PMID: 8906051.

59. Mørk NJ, Helsing P, Norvang LT. Do port wine stains recur after successful treatment with pulsed dye laser? *J Eur Acad Dermatol Venereol* 1998;11:S142-3. PMID: 10221131.

60. Michel S, Landthaler M, Hohenleutner U. Recurrence of port-wine stains after treatment with the flashlamp-pumped pulsed dye laser. *Br J Dermatol.* 200;143(6):1230-4. PMID: 11122026.

61. Huikeshoven M, Koster PH, de Borgie CA, *et al.* Redarkening of port-wine stains 10 years after pulsed-dye-laser treatment. *N Engl J Med.* 2007;356(12):1235-40. PMID: 17377161.

62. van der Horst CM, Koster PH, de Borgie CA, Bossuyt PM, van Gemert MJ. Effect of timing of treatment of port-wine stains with the flashlamp-pumped pulsed-dye laser. *N Engl J Med* 1998338:1028-33. PMID: 9535667.

63. Barsky SH, Rosen SH, Geer DE, Noe JM. The nature and evolution of port wine stains: a computer-assisted study. J Invest Dermatol 1980;74:154-7. PMID: 7359006.

64. Rosen S, Smoller BR. Port-wine stains: a new hypothesis. J Am Acad Dermatol 1987;17:164-6. PMID: 3611452.

65. Hohenleutner U. Hilbert M, Wlotzke U. Landthaler M. Epidermal damage and limited coagulation depth with the flashlamp-pumped dye laser: a histochemical study. J Invest Dermatol 1995;104:798-802. PMID: 7738359.

66. Nelson JS, Geronemus RG. Redarkening of Port-wine Stains 10 years after Laser treamtment (comment and author reply). *N Engl J Med* 2007;356(26):2745-6. PMID: 17596612.

Chapter 3

Ocular Manifestations of Sturge-Weber Syndrome

Tammy L. Yanovitch, M.D., M.HSc.

Sharon F. Freedman, M.D.

STURGE-WEBER SYNDROME IS A DISEASE characterized by congenital cephalic vascular malformations that affect the skin, eye, and central nervous system. This chapter summarizes the ocular findings of Sturge-Weber syndrome and reviews the treatment of these conditions.

OCULAR FINDINGS

In addition to the near universal finding of a port-wine stain, at least half of Sturge-Weber syndrome patients have one or more ocular findings. The most common ocular findings of Sturge-Weber syndrome are glaucoma and conjunctival, episcleral, and choroidal hemangiomas.[1,2,3] Other ocular findings include corneal enlargement, developmental angle anomaly, and iris heterochromia. **Table 3-1** summarizes the prevalence of ocular findings of individuals with Sturge-Weber syndrome.[4,5]

The most obvious sign of Sturge-Weber syndrome is a cutaneous capillary vascular malformation (port-wine stain) that typically affects the upper face (**Figure 3-1**).[6] Although it is classically unilateral, a port-wine stain may be bilateral and may involve the lower face and body; nevertheless, neurological manifestations of the syndrome rarely occur unless there is an upper facial nevus. The presence of a port-wine stain on either the upper or lower eyelids is associated with increased risk of glaucoma.[1,7]

Pulsed dye laser photocoagulation may be used to lighten the skin lesion.[8,9] This laser treatment theoretically injures blood vessels without damaging other skin constituents. More detailed information regarding the

Publication	Sharan % (n)	Sullivan % (n)	Gupta % (n)
Number of patients	41	51	62
Anterior Segment Findings			
Conjunctival/Episcleral Hemangioma	12.2 (5)	69.0 (35)	NR
Iris Heterochromia	NR	NR	14.5 (9)
Increased Corneal Diameter	36.6 (15)	51.0 (26)	NR
Developmental Angle Anomaly	66.7 (16)	26.0 (51)	NR
Glaucoma or Ocular Hypertension	58.5 (24)	71.0 (36)	35.4 (22)
Posterior Segment Findings			
Diffuse Choroidal Hemangioma	24.4 (10)	55.0 (28)	NR
Retinal Detachment	2.4 (1)	10.0 (5)	NR
Other			
Strabismus	0	14.0 (7)	NR

Table 3-1. Prevalence of ocular findings in Sturge-Weber syndrome patients. NR=not reported.

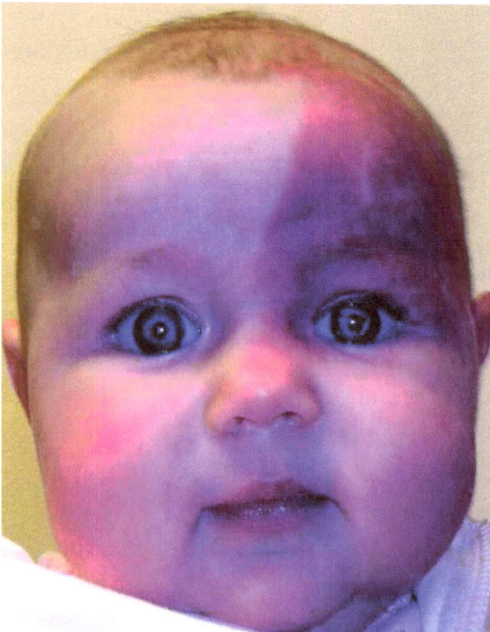

Figure 3-1. A photograph of an infant girl with a port-wine stain in the ophthalmic distribution of the trigeminal nerve on the left. She did not have evidence of glaucoma at this time. Note the symmetric corneal diameters and absence of episcleral vessel congestion.

treatment of port-wine stains is provided in Chapter 2. Although pulsed dye laser therapy could in theory lead to an increase in intraocular pressure, a recent study failed to find an association between pulsed dye laser treatment and glaucoma in a cohort of Sturge-Weber syndrome patients.[5,10]

In the anterior segment, corneal enlargement may occur.[4,5] The cornea is the transparent part of the eye that covers the iris, pupil, and anterior chamber. The average corneal diameter of a newborn infant is 10 mm; with glaucoma, the diameter increases to > 12 mm. This enlargement is often accompanied by buphthalmos (enlargement of the globe) secondary to Sturge-Weber syndrome-associated congenital glaucoma. **Figure 3-2** shows a photograph of an African boy with Sturge-Weber syndrome who has increased corneal diameters secondary to congenital glaucoma. Congenital glaucoma may cause tears (Haab stria) in the Descemet membrane of the cornea. Corneal edema and scarring may decrease the vision in the affected eye. The pathogenesis and treatment of glaucoma in Sturge-Weber syndrome is covered in more detail later in this chapter.

Additionally, developmental anomaly of the angle may occur.[11] The angle of the eye is the area at the base of the cornea that is responsible for draining the aqueous humor or fluid from the eye via the anterior chamber. If this fluid does not drain properly, pressure inside the eye increases and damages the optic nerve. Hence, angle anomalies are hypothesized to be one of the causative factors in Sturge-Weber syndrome-associated glaucoma.

Iris heterochromia (asymmetry in iris coloration) is present in 10% of patients with Sturge-Weber syndrome. The darker iris is located on the affected

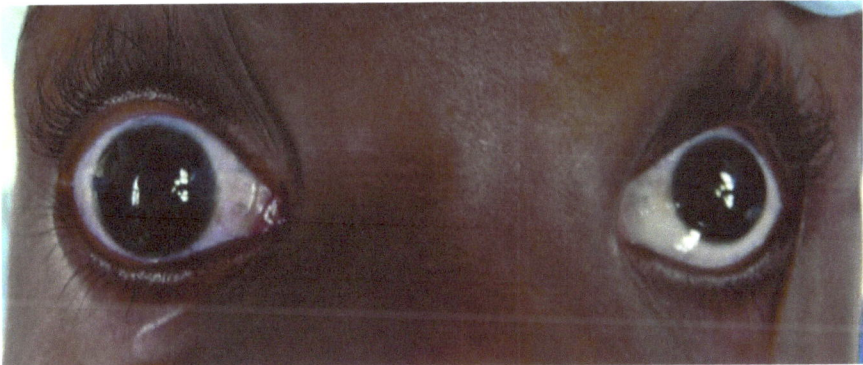

Figure 3-2. A picture showing a boy with Sturge-Weber syndrome from Africa who has untreated congenital onset glaucoma. Note the enlarged corneal diameter and conjunctival injection of the right eye compared with the left eye. His port wine stain is bilateral.

side, indicating an increase in melanocyte number or activity. Previous reports have suggested that the presence of iris heterochromia may signal an increased risk for glaucoma in Sturge-Weber syndrome patients.[12]

Conjunctival/episcleral hemangiomas are small, cavernous proliferations of blood vessels on the surface of the eye that occur in about 70% of Sturge-Weber syndrome patients. **Figure 3-3** demonstrates dilated capillaries on the surface of the conjunctiva of a patient with Sturge Weber syndrome. Conjunctival/episcleral hemangiomas are benign, but may result in ocular irritation. These lesions may be a harbinger of complications related to an underlying choroidal hemangioma, with associated increased risks of choroidal effusion or suprachoroidal hemorrhage during intraocular surgery.[13,14]

In addition to the conjunctiva and episclera, hemangiomas may also affect the posterior segment in patients with Sturge-Weber syndrome. Choroidal hemangiomas are congenital vascular hamartomas that occur in 20% to 50% of individuals with Sturge-Weber syndrome. These lesions are typically diffuse (**Figure 3-4**). The term "tomato catsup" fundus has been used to describe the reddish-orange appearance resulting from a diffuse choroidal hemangioma.[15]

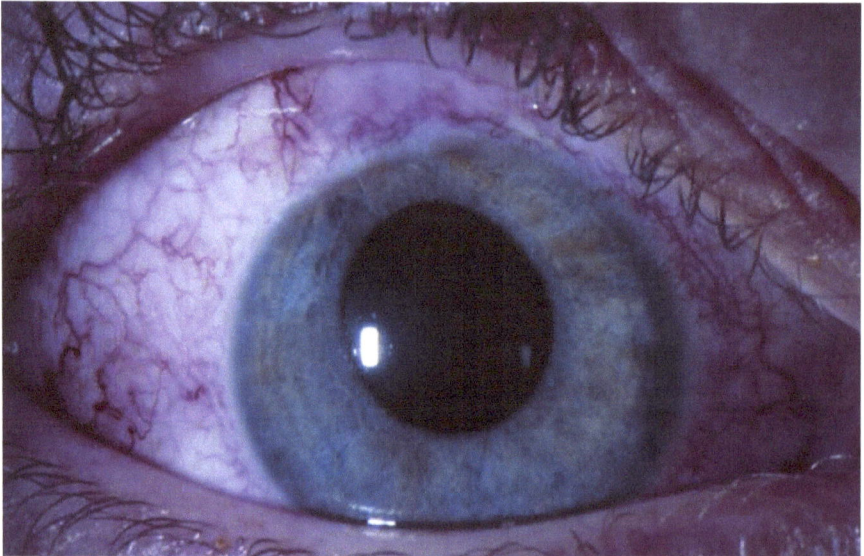

Figure 3-3. An image of a conjunctival/episcleral hemangioma in a patient with Sturge-Weber syndrome. The vessels appear dilated and tortuous. This eye also has glaucoma and has undergone successful aqueous drainage device surgery. The end of the short tube is present in the anterior chamber at 11:30 o'clock.

Choroidal hemangiomas are typically ipsilateral to the skin lesion. Like port-wine stains, the presence of a choroidal hemangioma increases the risk of glaucoma.

TREATMENT OF OCULAR COMPLICATIONS

For ophthalmologists, the main clinical challenges in caring for individuals with Sturge-Weber syndrome are related to the treatment of glaucoma and choroidal hemangiomas and their associated complications.

Glaucoma

In 1973, Weiss proposed his dual-origin hypothesis for glaucoma in Sturge-Weber syndrome. He argued that congenital glaucoma associated with Sturge-Weber syndrome was caused by both angle abnormalities and elevated episcleral venous pressure.[16] Histological observations support the presence of angle anomalies in individuals with Sturge-Weber syndrome who have congenital glaucoma, including: 1) poorly developed scleral spur; 2) thickened uvea and trabecular meshwork; 3) anterior displacement of the iris root, and; 4) attachment of the ciliary muscle directly to the trabecular meshwork. In contrast, Weiss attributed glaucoma in childhood and young adults to elevated episcleral venous pressure. This finding was supported by Phelps, who has observed that episcleral hemangiomas are present in eyes with glaucoma and that episcleral venous pressures are elevated in eyes with glaucoma.[3]

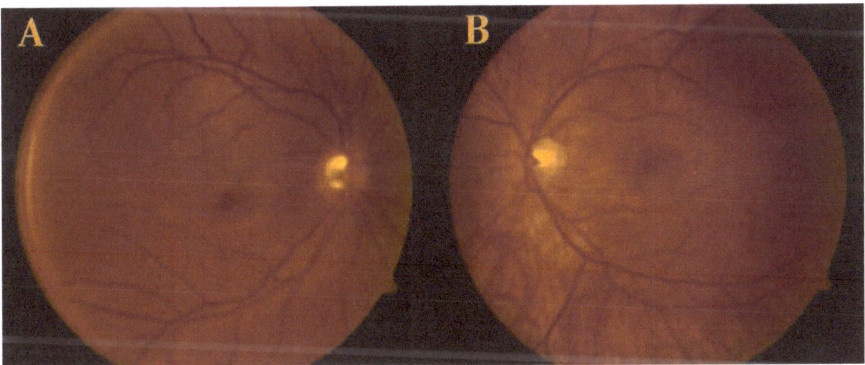

Figure 3-4A and B. Fundus photographs of a patient with Sturge-Weber syndrome. The right eye (4A) has a diffuse choroidal hemangioma. Contrast the reddish-orange color of the fundus in the right eye (4A) with that in the left eye (4B). The normal choroidal vasculature is hidden by the choroidal hemangioma in the right eye (4A), but is easily visualized in the left eye (4B).

Glaucoma occurs in 30% to 70% of individuals with Sturge-Weber syndrome. It is usually unilateral in those with a unilateral cutaneous lesion and typically occurs ipsilateral to the nevus. Glaucoma related to Sturge-Weber syndrome is congenital 60% to 70% of the time, while the other individuals usually present in childhood or early adulthood. Congenital glaucoma presents with the classic triad of photosensitivity, blepharospasm, and epiphora. Other signs include buphthalmos and corneal enlargement and clouding. Patients with later-onset glaucoma caused by Sturge-Weber syndrome may complain of blurred vision secondary to myopic shift or they may be identified while still asymptomatic. Periodic ophthalmologic exams are necessary to monitor these patients for the development of glaucoma (**Table 3-2**).

The physical findings are similar to those seen with other forms of glaucoma and include elevated intraocular pressure and optic nerve head cupping (**Figure 3-5**). Infants and young children (up to 3 years of age) may have increased corneal diameter, corneal edema, Haab striae, increased axial eye length, and myopia. Older children do not typically have corneal findings or buphthalmos but may have increased axial eye length and myopia.

The primary indication for treatment of Sturge-Weber syndrome associated glaucoma is uncontrolled intraocular pressure with progressive optic nerve cupping. The treatment of Sturge-Weber syndrome associated glaucoma depends largely on the age of onset. Congenital Sturge-Weber syndrome-associated glaucoma is typically treated with angle surgery, either goniotomy or trabeculotomy. The reported success rates for repeated goniotomy and/or trabeculotomy surgery in this setting is about 70% after four years of follow-up.[17] Combined trabeculotomy/trabeculectomy surgery is also used in congenital Sturge-Weber syndrome-associated glaucoma. This procedure is advocated because it addresses both issues underlying Sturge-Weber syndrome glaucoma—the elevated episcleral venous

Patient Age	Timing for Ophthalmologic Examination
Birth to 12 months	Birth and every 3 months thereafter (may require examination under anesthesia to obtain intraocular pressure measurements)
12 months to 24 months	Every 6 months*
> 24 months	Annually*

Table 3-2. Screening recommendations for glaucoma in Sturge-Weber syndrome patients. *These visits should occur more frequently if glaucoma has already been diagnosed.

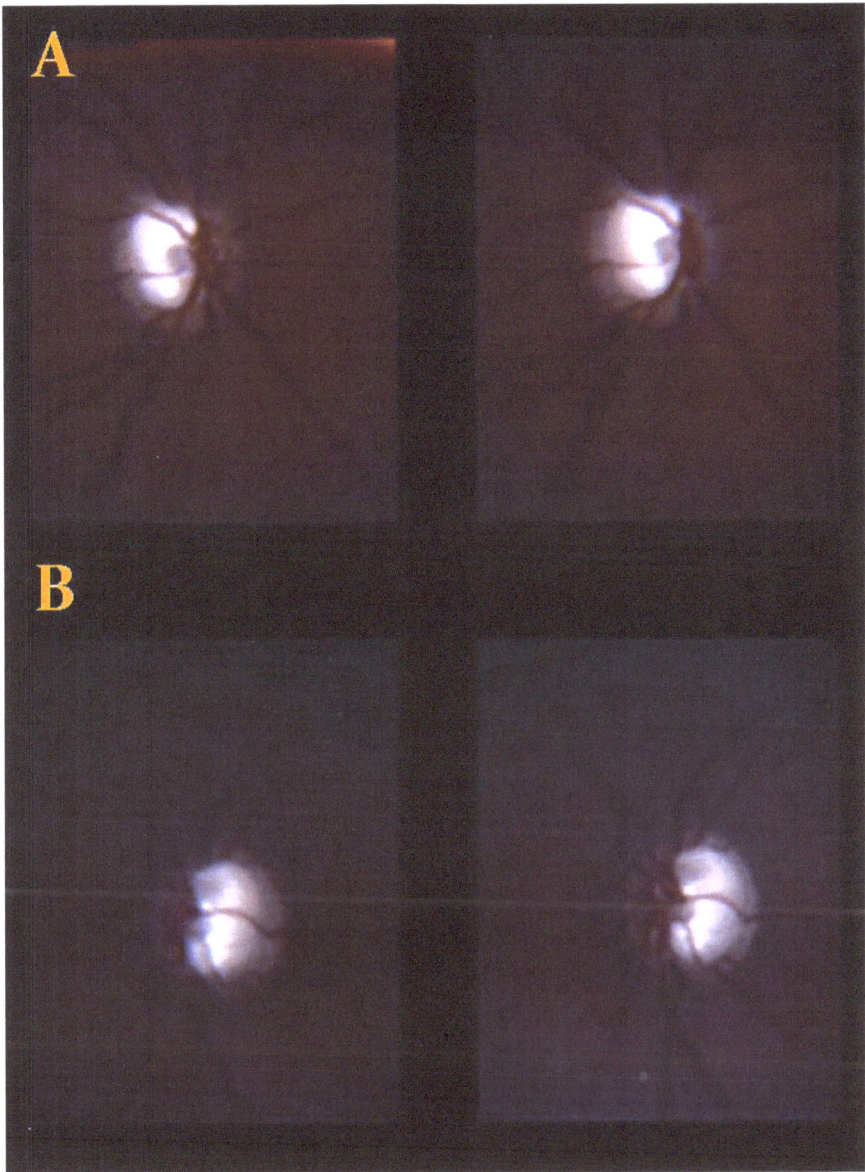

Figure 3-5A and B. Stereographic optic nerve head photos in a young girl with Sturge-Weber syndrome associated glaucoma. She has extensive optic nerve head cupping and diffuse choroidal hemagiomas in both eyes. Her port-wine stain is bilateral. She presented with congenital glaucoma and underwent angle surgery as a baby. After this procedure, her pressures remained adequately controlled until she was 7 years old, when she required aqueous drainage device implantation.

pressure and the drainage angle abnormalities. The reported success rate at 24 months for the combined procedure is high. [18,19]

Initially, most clinicians treat childhood or adult-onset glaucoma due to Sturge-Weber syndrome with medical therapy. Aqueous suppressants, such as beta blockers and carbonic anhydrase inhibitors, are the most effective. Although prostaglandin analogues lower intraocular pressure in Sturge-Weber syndrome patients, several cases of choroidal effusions associated with the use of these medications have been reported. It is thought that the elevated episcleral venous pressure seen in Sturge-Weber syndrome patients and the enhanced uveoscleral outflow caused by the prostaglandin analogues leads to congestion of aqueous humor in the supracilliary choroidal space. This congestion results in a ciliochoroidal effusion. In general, more than half of the later-onset cases will ultimately fail medical therapy and require filtration or aqueous drainage device surgery.[20]

Filtration surgery, aqueous drainage device implantation, and/or ciliary body destruction are used for the treatment of both refractory congenital and later-onset glaucoma. Both trabeculectomy and non-penetrating deep sclerectomy have been reported to lower intraocular pressure effectively in Sturge-Weber syndrome patients.[21,22] Small case series have been published on the Molteno, Ahmed and Baerveldt glaucoma drainage devices for treatment of refractory Sturge-Weber syndrome associated glaucoma.[23,24,25] Although all of the published aqueous drainage device series had limited follow-up data, the two-stage Baerveldt showed the most promising results, with a complete success rate at 35 months. Other experts have recommended reduction of aqueous production by destruction of the ciliary body.[26] Because of the rarity of the condition and high rate of complications, the optimal surgical method of lowering intraocular pressure in Sturge-Weber syndrome with refractory or later-onset glaucoma remains uncertain.

Filtration surgery and other intraocular procedures to control glaucoma in patients with Sturge-Weber syndrome are sometimes complicated by perioperative choroidal effusion or expulsive choroidal hemorrhage. Effusions are thought to occur due to dramatic intraoperative changes in intraocular pressure.[13,14] As previously mentioned, patients with Sturge-Weber syndrome have increased episcleral venous pressure with resulting elevation in venous pressure within the ciliary body and choroid. During surgery, the intraocular pressure falls rapidly and fluid transudates from the intravascular to extravascular space. Similarly, hemorrhage can also result from rupture of fragile vessels due to a sudden intra-operative or post-operative decrease in intraocular pressure.

Prophylactic measures to prevent choroidal effusion and hemorrhage include posterior sclerotomy, radiotherapy to choroidal hemangiomas, and electrocautery of the anterior episcleral vascular anomaly.[27,28] The need for these prophylactic measures is debated, but most surgeons agree that it is important to avoid intraoperative and post-operative hypotony in these patients. In Sturge-Weber syndrome patients requiring glaucoma surgery, the authors prefer the Baerveldt aqueous drainage device, either in two stages or with double, complete ligation of the tubing anterior to the reservoir, without use of venting slits and with use of an anterior chamber maintainer during tube placement to avoid the need for viscoelastic that may result in an unwanted, extremely high intraocular pressure post-operatively. In eyes with limited visual potential or at particularly high risk for choroidal hemorrhage, limited transscleral diode laser cycloablation may be performed.

In Sturge-Weber syndrome patients with refractory glaucoma, control of intraocular pressure may be lost and complications related to glaucoma surgery may occur at any time. In addition, infants and children with Sturge-Weber syndrome-associated glaucoma may also develop secondary ophthalmologic issues such as amblyopia, refractive error, corneal clouding/scarring, strabismus, and cataract. Hence, it is important for these patients to receive a lifetime of close ophthalmologic follow-up.

Choroidal Hemangioma

As many as half of Sturge-Weber syndrome patients have choroidal hemangiomas. Choroidal hemangiomas are benign, vascular hamartomas that appear as reddish-orange choroidal thickening. These lesions may have focal areas of leaky vessels that lead to vision loss secondary to serous retinal detachment. Chronic changes related to serous retinal detachment include fibrous metaplasia, atrophy of the retinal pigment epithelium, and cystoid macular edema. Management options for vision-threatening, leaking choroidal hemangiomas include photocoagulation, diathermy, cryotherapy, external irradiation, and brachytherapy, with the treatment goal being resolution of subretinal fluid. [29,30,31,32]

Ancillary testing for choroidal hemangiomas includes ultrasonography, flourescein angiography, indocyanine green angiography, and magnetic resonance imaging. With A-scan ultrasonography, choroidal hemagiomas appear to have high internal reflectivity.[33] These lesions are acoustically solid on B-scan ultrasound (**Figure 3-6**). On intravenous flourescein angiography, choroidal hemangiomas show mild hyperintensity during the pre-arterial and arterial phases, moderate hyperintensity during the venous phase, and extreme hyperintensity during the late phase (**Figure 3-7**).[34] With indocyanine green angiography, they appear extremely hyperintense at 1 minute and moderately hyperintense at 8 and 20 minutes.[35,36] Choroidal hemangiomas demonstrate hyperintensity on T1 weighted and T1 with gadolinium and hyperintensity/ isointensity on T2 weighted MRI.[37]

In addition to diagnosing choroidal hemangioma, ultrasonography and angiography assist in the diagnosis of associated serous retinal detachments associated with choroidal hemangiomas. Optical coherence tomography allows for the visualization of retinal detachment and sub-retinal fluid. These methods are also useful in following patients after undergoing therapy for serous retinal detachments.

Serious retinal detachments associated with diffuse choroidal hemangioma in Sturge-Weber syndrome patients have been successfully treated with radiation and photodynamic therapy.[29, 30, 31, 32] Several case reports suggest that these treatments are effective, but there are no comparative studies. Complications of

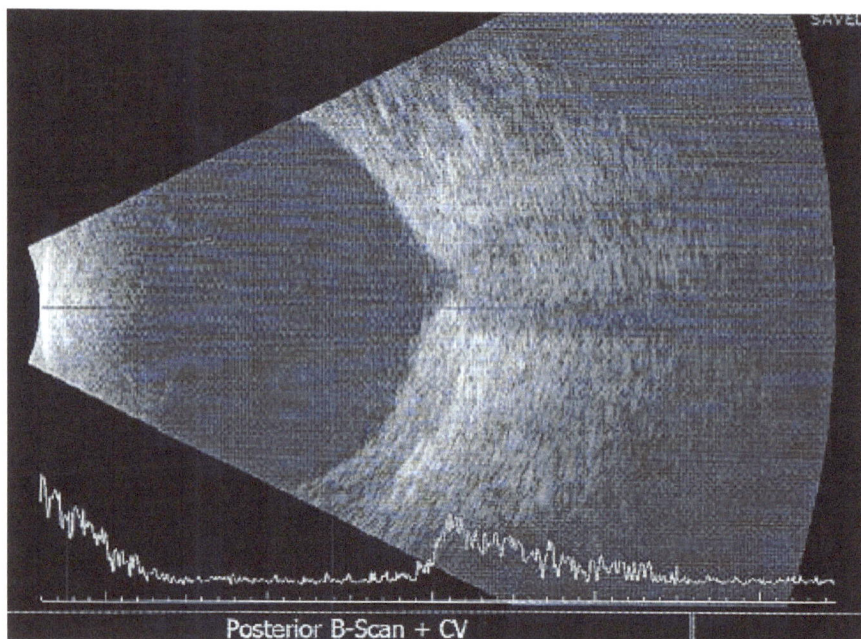

Figure 3-6. B-scan ultrasound image demonstrating marked thickening of the choroid found in a Sturge-Weber syndrome patient with a diffuse choroidal hemangioma. Note the large optic nerve cup also visible on this scan. This B-scan corresponds to the left eye of the nerve photos shown in Figure 3-5.

radiation therapy include retinopathy, cataract, and optic neuropathy. In contrast, photodynamic therapy may lead to pigmentary scarring of the retina.

One individual with Sturge-Weber syndrome had a refractory serous retinal detachment that resolved following treatment with an anti-VEGF agent.[38]

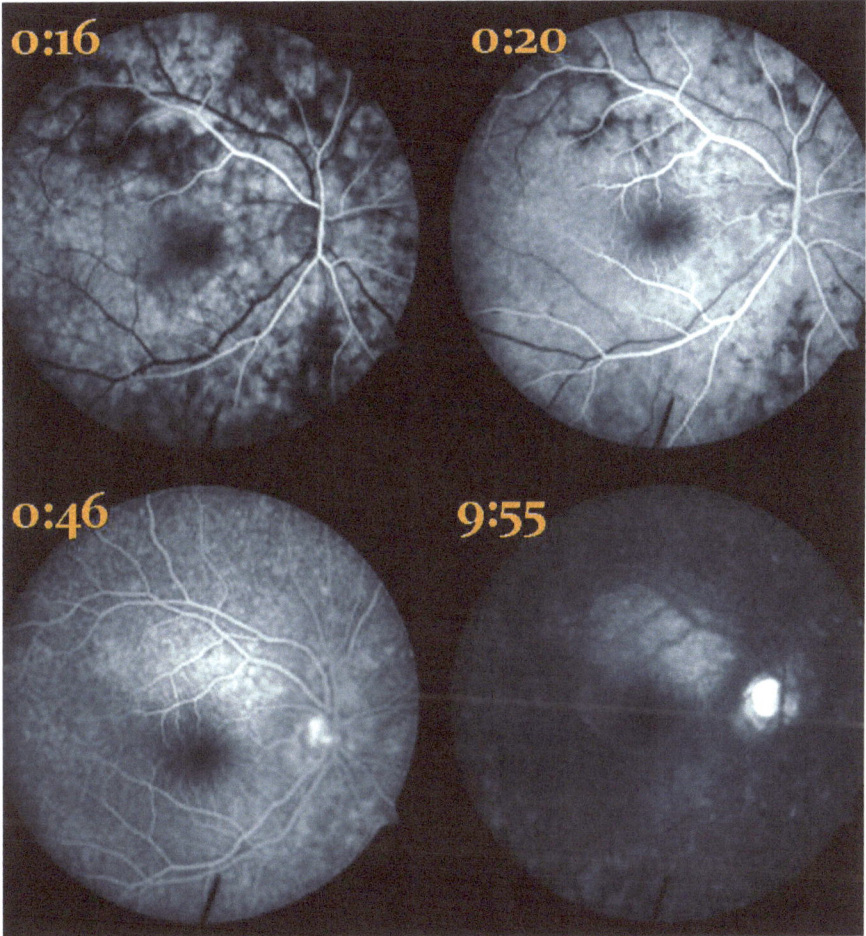

Figure 3-7. Fluorescein angiogram of the right eye in a Sturge-Weber syndrome patient with a diffuse choroidal hemangioma and a focal area of sub-retinal fluid. The patient was a 17-year-old boy who had noted recent onset of blurred vision in his right eye. He also complained of metamorphopsia in his central visual field. There is early hyperfluorescence along the supertemporal arcade with associated leakage and serous pooling in the foveal space. The serous detachment was successfully treated with photodynamic therapy and the patient's visual acuity returned to baseline.

However, the underlying mechanism of action and the long-term implications of anti-VEGF therapy in this scenario remain unclear.

In infants and children, choroidal hemangiomas may also result in anisometropia (with increased hyperopia in the affected eye) and amblyopia. Individuals with Sturge-Weber syndrome with a choroidal hemangioma require lifelong surveillance to detect enlargement and/or leakage.

Conclusions

In conclusion, Sturge-Weber syndrome is a rare, sporadic disease that occurs in 1:20,000-50,000 live births. The most serious ophthalmologic complications of Sturge-Weber syndrome are glaucoma and choroidal hemangioma. Both of these conditions require prompt recognition and early treatment to prevent vision loss, so it is imperative that these individuals undergo regular ophthalmologic examinations and receive prompt treatment. Given the potential loss of vision in an affected eye, care should be taken to protect the uninvolved eye from injury and other diseases.

REFERENCES

1. Stevenson RF, Thomson HG, Morin LD. Unrecognized ocular problems associated with port-wine stain of the face in children. *Can Med Assoc J* 1974;111:953-954.

2. Witschel H, Font RL. Hemangioma of the choroid. A clinicopathologic study of 71 cases and a review of the literature. *Surv Ophthalmol* 1976;20:415-431.

3. Phelps CD. The pathogenesis of glaucoma in Sturge-Weber syndrome. *Ophthalmology* 1978;85:276-286.

4. Sullivan TJ, Clarke MP, Morin JD. The ocular manifestations of the Sturge-Weber syndrome. *J Pediatr Ophthal Strabismus* 1992;29:349-356.

5. Sharan S, Swamy B, Taranath DA et al. Port-wine vascular malformations and glaucoma risk in Sturge-Weber syndrome. *J AAPOS* 2009;13:374-378.

6. Alexander GL, Norman RM. *Sturge-Weber Syndrome*. Bristol: John Wright & Sons Ltd, 1960.

7. Hamush NG, Coleman AL, Wilson MR. Ahmed glaucoma valve implant for management of glaucoma in Sturge-Weber syndrome. *Am J Ophthalmol* 1999;128:758-760.

8. Tan OT, Sherwood K, Gilchrest BA. Treatment of children with port-wine stains using the flashlamp-pulsed tunable dye laser. *N Engl J Med* 1989;320:416-421.

9. Stier MF, Glick SA, Hirsch RJ. Laser treatment of pediatric vascular lesions: Port wine stains and hemangiomas. *J Am Acad Dermatol* 2008;58:261-285.

10. Ray D, Mandal AK, Chandrasekhar G, Naik M, Dhepe N. Port-wine vascular malformations and glaucoma risk in Sturge-Weber syndrome. *J AAPOS* 2010;14:105.

11. Cibis GW, Tripathi RC, Tripathi BJ. Glaucoma in Sturge-Weber syndrome. *Ophthalmology* 1984;91:1061-1071.

12. Aggarwal NK. Allowing independent forensic evaluations for Guantanamo detainees. *J Am Acad Psychiatry Law* 2009;37:533-537.

13. Bellows AR, Chylack LT, Jr., Epstein DL, Hutchinson BT. Choroidal effusion during glaucoma surgery in patients with prominent episcleral vessels. *Arch Ophthalmol* 1979;97:493-497.

14. Christensen GR, Records RE. Glaucoma and expulsive hemorrhage mechanisms in the Sturge- Weber syndrome. *Ophthalmology* 1979;86:1360-1364.

15. Susac JO, Smith JL, Scelfo RJ. The "tomatoe-catsup" fundus in Sturge-Weber syndrome. *Arch Ophthalmol* 1974;92:69-70.

16. Weiss DI. Dual origin of glaucoma in encephalotrigeminal haemangiomatosis. *Trans Ophthalmol Soc U K* 1973;93:477-493.

17. Olsen KE, Huang AS, Wright MM. The efficacy of goniotomy/trabeculotomy in early-onset glaucoma associated with the Sturge-Weber syndrome. *J AAPOS* 1998;2:365-368.

18. Mandal AK. Primary combined trabeculotomy-trabeculectomy for early-onset glaucoma in Sturge-Weber syndrome. *Ophthalmology* 1999;106:1621-1627.

19. Luntz MH. Trabeculotomy-trabeculectomy for early-onset glaucoma. *Ophthalmology* 2000;107:624-625.

20. Iwach AG, Hoskins HD, Jr., Hetherington J, Jr., Shaffer RN. Analysis of surgical and medical management of glaucoma in Sturge-Weber syndrome. *Ophthalmology* 1990;97:904-909.

21. Ali MA, Fahmy IA, Spaeth GL. Trabeculectomy for glaucoma associated with Sturge-Weber syndrome. *Ophthalmic Surg* 1990;21:352-5XX63X.

22. Audren F, Abitbol O, Dureau P et al. Non-penetrating deep sclerectomy for glaucoma associated with Sturge-Weber syndrome. *Acta Ophthalmol Scand* 2006;84:656-660.

23. Awad AH, Mullaney PB, Al-Mesfer S, Zwaan JT. Glaucoma in Sturge-Weber syndrome. *J AAPOS* 1999;3:40-45.

24. Budenz DL, Sakamoto D, Eliezer R, Varma R, Heuer DK. Two-staged Baerveldt glaucoma implant for childhood glaucoma associated with Sturge-Weber syndrome. *Ophthalmology* 2000;107:2105-2110.

25. Amini H, Razeghinejad MR, Esfandiarpour B. Primary single-plate Molteno tube implantation for management of glaucoma in children with Sturge-Weber syndrome. *Int Ophthalmol* 2007;27:345-350.

26. van EC, Goethals M, Dralands L, Casteels I. Treatment of glaucoma in children with Sturge-Weber syndrome. *J Pediatr Ophthalmol Strabismus* 2000;37:29-34.

27. Eibschitz-Tsimhoni M, Lichter PR, Del Monte MA et al. Assessing the need for posterior sclerotomy at the time of filtering surgery in patients with Sturge-Weber syndrome. *Ophthalmology* 2003;110:1361-1363.

28. Mandal AK, Gupta N. Patients with Sturge-Weber syndrome. *Ophthalmology* 2004;111:606.

29. Zografos L, Bercher L, Chamot L, Gailloud C, Raimondi S, Egger E. Cobalt-60 treatment of choroidal hemangiomas. *Am J Ophthalmol* 1996;121:190-199.

30. Zografos L, Egger E, Bercher L, Chamot L, Munkel G. Proton beam irradiation of choroidal hemangiomas. *Am J Ophthalmol* 1998;126:261-268.

31. Grant LW, Anderson C, Macklis RM, Singh AD. Low dose irradiation for diffuse choroidal hemangioma. *Ophthalmic Genet* 2008;29:186-188.

32. Anand R. Photodynamic therapy for diffuse choroidal hemangioma associated with Sturge Weber syndrome. *Am J Ophthalmol* 2003;136:758-760.

33. Gitter KA, Meyer D, Sarin LK, Keeney AH, Justice J, Jr. Fluorescein and ultrasound in diagnosis of intraocular tumors. *Am J Ophthalmol* 1968;66:719-731.

34. Horgan N, O'Keefe M, McLoone E, Lanigan B. Fundus fluorescein angiographic characterization of diffuse choroidal hemangiomas. *J Pediatr Ophthalmol Strabismus* 2008;45:26-30.

35. Schalenbourg A, Piguet B, Zografos L. Indocyanine green angiographic findings in choroidal hemangiomas: A study of 75 cases. *Ophthalmologica* 2000;214:246-252.

36. Wen F, Wu D. Indocyanine green angiographic findings in diffuse choroidal hemangioma associated with Sturge-Weber syndrome. *Graefes Arch Clin Exp Ophthalmol* 2000;238:625-627.

37. Griffiths PD, Boodram MB, Blaser S et al. Abnormal ocular enhancement in Sturge-Weber syndrome: correlation of ocular MR and

CT findings with clinical and intracranial findings. *AJNR Am J Neuroradiol* 1996;17:749-754.

38. Paulus YM, Jain A, Moshfeghi DM. Resolution of persistent exudative retinal detachment in a case of Sturge-Weber syndrome with anti-VEGF administration. *Ocul Immunol Inflamm* 2009;17:292-294.

Chapter 4

Neurological Manifestations of Sturge-Weber Syndrome

Anne M. Comi, M.D.

E. Steve Roach, M.D.

John B. Bodensteiner, M.D.

EPILEPSY, MENTAL RETARDATION, AND FOCAL NEUROLOGICAL deficits are the major neurologic abnormalities of Sturge-Weber syndrome.[1,2] Most patients with radiographic evidence of an intracranial angioma develop seizures, but not all patients are mentally retarded.[3-5] Children without frank mental retardation commonly have learning disability, attention deficit, or behavioral disturbance. The severity of neurologic impairment varies, however, and patients with a nevus and seizures who have normal intelligence and lack focal neurologic deficits are common.

The age when symptoms appear varies, as does the overall clinical severity, but early seizure onset correlates loosely with the risk of both future mental retardation and refractory epilepsy. The clinical condition eventually stabilizes in most patients, often leaving hemiparesis, hemianopia, cognitive impairment, or epilepsy, but without further deterioration. Sturge-Weber is an enigmatic syndrome that is seldom hard to recognize but often difficult to predict or treat effectively. This difficulty is due to the highly individual nature of the clinical features and to the lack of uniformly effective treatment for some of the more serious complications.

NEUROLOGIC COMPLICATIONS

Nevus Distribution and Neurological Impairment

The classic port-wine nevus of Sturge-Weber syndrome involves the forehead and upper eyelid region (see Chapter 2), but the lesion often extends onto both sides of the face and even onto the trunk and extremities. Patients whose nevus involves only the trunk or the lower face do not usually have a brain angioma and thus have little risk of developing neurologic problems.[6-8] Although the cutaneous lesion is normally apparent at birth, only 10% to 20% of children with a forehead port-wine nevus have a leptomeningeal angioma.[7] Children without brain involvement are not usually included among Sturge-Weber patients, but such children need to be followed closely until the absence of brain dysfunction is assured. Occasionally children have the characteristic neurologic and radiographic features of Sturge-Weber syndrome but have no skin abnormality.[9-12]

The leptomeningeal angioma is typically on the same side as the facial nevus, but bilateral brain lesions occur in at least 15% of patients, including some with a unilateral cutaneous nevus.[13] In occasional patients, the cerebral angioma is contralateral to the facial nevus.[14] The extent of the nevus correlates poorly with the severity of neurologic impairment.[6] However, patients with extensive facial nevi are probably more likely to have bilateral brain angiomas, and children with bilateral brain lesions in turn have a greater risk of neurologic impairment and earlier seizure onset.[8,15]

Epileptic Seizures

Seizures typically begin acutely at several months of age, often in conjunction with hemiparesis or other focal deficits. Children whose seizures begin prior to two years of age and those whose seizures can not be controlled with medication are more likely to become mentally retarded. In contrast, patients without seizures are usually not mentally retarded, and few patients who remain normal past age three are destined to have severe intellectual impairment.

Seizures occur in 72% to 80% of Sturge-Weber patients with unilateral brain lesions and in 93% of patients with bihemispheric involvement.[15,16] The risk of developing seizures is highest in the first two years and then tends to decrease: 75% begin by one year of age, 86% by age two, and 95% by age five.[3] As a rule individuals who have frequent seizures, particularly at a very young age, have a greater likelihood of mental retardation.[17] Focal motor seizures or generalized tonic-clonic seizures are common initially, but infantile spasms, myoclonic seizures, and atonic seizures occur.[18-23] Older children and adults are more likely to have complex partial seizures or focal motor seizures. In one series of 77 patients with Sturge-Weber syndrome and seizures, 14% had generalized seizures (in addition to complex partial or focal seizures) that were exacerbated by carbamazepine or oxcarbazine and which responded to valproic acid.[24] Some individuals experience clusters of intense seizure activity interspersed with long seizure-free intervals even without medication, while others have frequent or prolonged seizures despite high doses of multiple medications.[19,24]

Cognitive and Psychological Impairment

Developmental milestones are usually attained on schedule during the first few months of life, but mental deficiency eventually develops in about half of the children with Sturge-Weber syndrome.[25] Only 8% of the patients with bilateral brain involvement are cognitively normal.[15] The severity of intellectual impairment ranges from mild to profound, and patients with normal intelligence sometimes develop social, psychological, or behavioral problems.[26] Patients who do not have seizures usually develop normally and those whose seizures are fully controlled are more likely to function normally. While continued seizures no doubt contribute to poor mental function, the occurrence of cerebral dysplasia near the vascular lesion could play a role in both intractable epilepsy and faulty mental development.[27] Most patients with Sturge-Weber syndrome take antiepileptic drugs, and to some extent, these medications can impair cognition and alter behavior.

Behavioral abnormalities and learning disabilities can be problematic even in patients who are not mentally retarded, and of course some children with intellectual impairment also have coexisting attentional and behavioral disturbances (see Chapter 5). Attention deficit disorder without hyperactivity is a particular concern, because poor school performance in this setting is all too easily attributed to lack of intelligence due to the syndrome. Attention problems are responsive to stimulant medications, and these agents should be considered in individuals with significant dysfunction due to poor attention. Neuropsychological testing can help to identify at-risk children. Aggression toward others, defiant and oppositional actions, and self-abusive behavior are probably the most common behavioral concerns.

Depression is probably as common among older children and adolescents as it is in adults. Occasional patients develop frank psychosis.[28] Adults with Sturge-Weber syndrome and normal intelligence are often depressed. Half of the adults with Sturge-Weber syndrome in the Sujansky and Conradi survey had at some point been depressed, although not all of these people had sought treatment for depression.[4] Some respondents blamed their depression on frustration caused by poor seizure control, while others credited headaches or the cosmetic effect of the facial nevus. Depression is less common in intellectually impaired adults (38%), but these individuals may be harder to assess. In addition, intellectually handicapped adults sometimes develop aggression toward others and self-abusive tendencies.[4]

Focal Neurological Deficits

Sturge-Weber syndrome causes various focal neurological signs, but these deficits occur less consistently than epileptic seizures. Up to two thirds of the adults with Sturge-Weber syndrome have some type of focal neurological deficit.[4] The exact clinical deficiency depends on the extent of the intracranial vascular lesion and its location in the brain, but the deficit most often noted is hemiparesis. Because the occipital region is often involved, visual field defects are also common.

Hemiparesis often develops acutely in conjunction with the initial flurry of seizures. Seizure onset at less than six months of age has been related to increased severity of hemiparesis in a cohort of 77 patients seen at one center.[24] Although the weakness is often attributed to the seizures, hemiparesis may be permanent or persist much longer than the few hours typical of a postictal deficit. In children with both hemiparesis and seizures, it can be difficult to establish which came first.[29]

Neurological deterioration in these individuals sometimes occurs in an episodic, stepwise manner, with the trend overall toward steadily accumulating deficit.[30] Some children develop acute weakness without seizures, either as repeated episodes of weakness similar to transient ischemic attacks or as a single stroke-like episode with persistent deficit.[30] Children with early-onset persistent hemiparesis often exhibit impaired growth of the affected extremities, resulting in hemiatrophy.

Minor head injuries such as from falls in toddlers or young children with Sturge-Weber can trigger increased seizures or stroke-like episodes, some of which are associated with vomiting and headache.[31]

Macrocephaly

Cranial enlargement has been documented in several children with Sturge-Weber syndrome, and a few of these children subsequently developed progressive hydrocephalus.[32-35] Head enlargement in some children seems to stem from obstruction of the major venous drainage system. Others have increased venous pressure due to increased cerebral blood flow secondary to the meningeal angioma. Shapiro and Shulman shunted two children with hydrocephalus and increased intracranial pressure due to absent jugular outflow.[34] The child with progressive head enlargement and hydrocephalus reported by Fishman and Baram had no drainage through the vein of Galen and secondary dilatation of the superficial cortical veins.[32] In some patients the ventricular enlargement stops even without intervention.[36]

Headache

Headache is common among Sturge-Weber patients, but whether they have a higher headache frequency than the population as a whole has not been resolved. It can be difficult to distinguish complicated migraine from episodic neurological deficits generated by the syndrome itself.

One survey estimated that 44% of Sturge-Weber patients have headaches, more than double the rate of headaches in the general populace.[37] However, only 71 of 500 questionnaires distributed in this study were returned, and the rate of survey completion might be substantially higher for patients with headache than for individuals without headache. In this group, migraine was the most frequent headache type. Headache commonly occurred just after an epileptic seizure, and a few patients' headaches were attributed to glaucoma.[37] In another survey of adult Sturge-Weber patients (identified from the same patient database), 28 of 45 (62%) adult patients reported headache. The headache frequency could be ascertained in only 23 people, but many of these had frequent headaches: nine (39.1%) had daily headaches and four (17.4%) complained of headache once or twice per week.[4]

Headaches are common in the general population, so some patients with Sturge-Weber syndrome would be expected to develop headaches with or without the disorder. Headaches tend to begin at a younger age in individuals with Sturge-Weber syndrome than in other people.[38] Several additional factors could cause headache in an individual with Sturge-Weber syndrome. Glaucoma can cause head pain, but only a few of those reporting headache have glaucoma. Half of the adults with headaches acknowledge having depression, which can promote chronic, often daily, headaches. Many people in the surveys have headache just after an epileptic seizure, a common pattern in epileptic patients in general.

Autoregulation of cerebral blood flow is impaired in individuals with Sturge-Weber syndrome, even in the areas of the brain distant from the cerebral angioma.[39,40] Altered blood flow reactivity could increase the likelihood of headaches by simulating the pattern of abnormal vasomotor reactivity seen with migraine. Headache may also be associated with ischemia of the pain-sensitive meningeal structures surrounding the angioma. One adult with Sturge-Weber syndrome and an intractable headache with a prolonged visual aura demonstrated focal hyperemia on SPECT, increased occipital sulcal enhancement on contrast-enhanced magnetic resonance imaging (MRI), and occipital retention of contrast on computed tomographic angiography, findings that resolved after the headache stopped.[41]

Intracranial Hemorrhage

Although hemorrhage might be expected to result from an extensive intracranial venous anomaly, intracranial hemorrhage is in fact quite rare in individuals with Sturge-Weber syndrome. Harvey Cushing in 1906 described three patients with possible spontaneous intracranial hemorrhage.[42] However, all three of these patients developed acute weakness and seizures, a pattern seen in many children with Sturge-Weber syndrome who do not have hemorrhage, and he found no specific indication of hemorrhage at the autopsy done on two of these patients. At least one teenager with Sturge-Weber syndrome developed an intraparenchymal hemorrhage, which was later evacuated.[43] Another adult with Sturge-Weber syndrome had a subarachnoid hemorrhage.[44] Microscopic hemorrhages are sometimes demonstrated in autopsy or surgical specimens, but these seem to have little clinical significance.

Occasional patients have an actual arteriovenous malformation and others have extensively dilated veins. These patients might be expected to have a greater risk of hemorrhage than other Sturge-Weber patients, but individuals with such lesions are uncommon and there are few recorded examples of hemorrhage even in this group.

Mechanisms of Neurologic Deterioration

The extent and location of cerebral angiomatosis contributes to the neurologic outcome because children with an extensive vascular lesion often have more difficult seizures and more intellectual impairment. However, even children a large vascular lesion usually have normal neurologic function initially, and several factors, singly or in combination, promote neurological deterioration of these patients.

Numerous epileptic seizures no doubt contribute to the deterioration in some children. Certainly children with refractory seizures attributed to other causes often improve significantly once their seizures are controlled.[45] Chronic hypoxia of the cerebral cortex adjacent to the angioma resulting from reduced blood flow is associated with calcium deposition in the affected tissue (a phenomenon known as dystrophic calcification). Increased metabolic requirements during seizures could potentiate an oxygen deficit.[25] In addition, ictal single photon emission computed tomography (SPECT) studies in individuals with Sturge-Weber syndrome suggest that cerebral blood flow during seizures, rather than increasing bilaterally, either increases marginally or decreases to ischemic levels in involved or remote brain regions.[46,47]

Some children undergo saltatory decline via a series of distinct episodes of neurologic dysfunction, and episodic neurologic deficits can occur even without overt seizure activity.[30] Repeated venous occlusions may account for the step-wise deterioration in these children and the episodic neurologic dysfunction seen in other children without seizures.[30] Venous occlusion could also explain the typical first episode of neurologic dysfunction: the clinical picture at the time of the initial deterioration resembles the pattern seen with venous thromboses from other causes.

DIAGNOSIS OF STURGE-WEBER SYNDROME

Most children with a facial port-wine nevus do not have an intracranial angioma, and thus have no neurological dysfunction.[7] Neuroimaging studies and functional imaging tests help to distinguish the children with Sturge-Weber syndrome from those with an isolated cutaneous lesion. Neuroimaging, electroencephalography, and functional brain imaging with functional magnetic resonance imaging, positron emission tomography (PET) and SPECT may also help to define the extent of the intracranial lesion for possible epilepsy surgery.[45,48,49] Imaging techniques used to evaluate Sturge-Weber syndrome are discussed more thoroughly in Chapter 6.

Although gyral calcification is a classic feature of Sturge-Weber syndrome, the "trolley track" appearance this creates is not always present, especially just after birth. Calcification may become more apparent with age, and calcium can be demonstrated much more reliably with computed cranial tomography (**Figure 4-1**) than with either MRI or standard skull X-rays.[50,51]

Extensive cerebral atrophy is apparent even with computed tomography, but subtle atrophy is more readily demonstrated by magnetic resonance imaging.[50,52] Magnetic resonance imaging with gadolinium contrast (**Figure 4-2**) effectively demonstrates the abnormal intracranial vessels of Sturge-Weber syndrome, and gadolinium contrast MRI is currently the most accurate, widely available test to demonstrate intracranial involvement.[53-56] Post-contrast FLAIR and

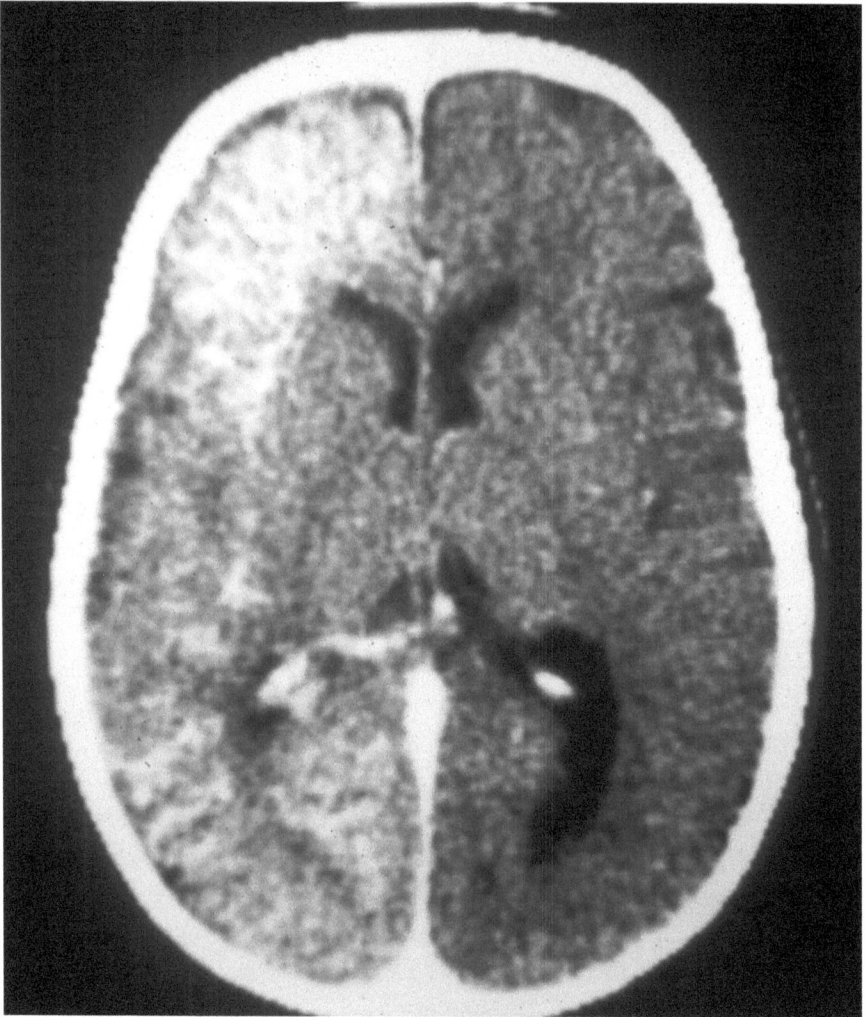

Figure 4-1. A: Computed cranial tomography from a typical patient with Sturge-Weber syndrome with extensive involvement of the right hemisphere.

susceptibility-weighted MR imaging may provide additional elucidation of involved brain areas.[57,58]

Catheter arteriography is no longer a routine part of the evaluation of Sturge-Weber syndrome, but it may be useful for patients with atypical features, or prior to surgery for epilepsy.[30] The superficial cortical veins are sparse (**Figure 4-3 A and B**), whereas the deep subependymal and medullary veins are enlarged and tortuous.[59,60] The sagittal sinus may not opacify after ipsilateral carotid injection, probably due to elimination of the superficial cortical veins.[61] The enlarged deep

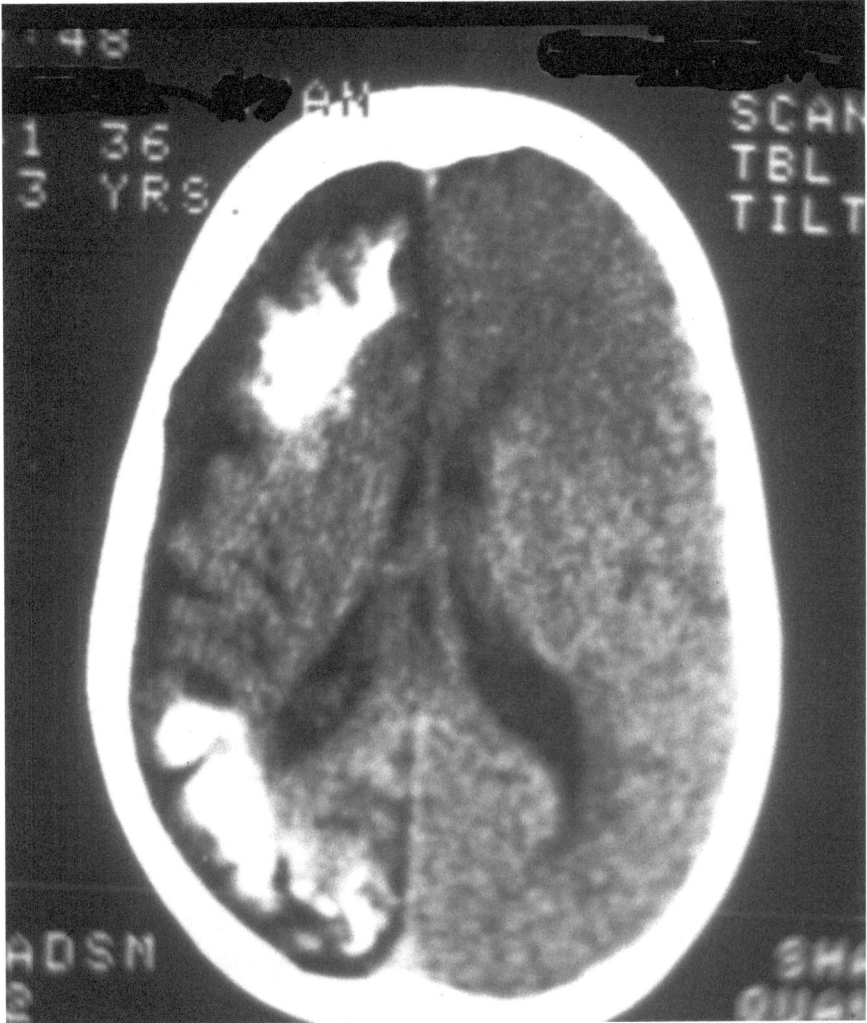

Figure 4-1. B: The same patient years later as atrophy and dystrophic calcification of the hemisphere has progressed.

venous channels may have a similar origin as they form collateral conduits for nonfunctioning cortical veins.[61] A homogeneous blush from the leptomeningeal angioma is sometimes visible. [62,63]

Functional imaging with PET demonstrates reduced metabolism of the brain adjacent to the leptomeningeal lesion, often extending well beyond the area of abnormality depicted by computed tomography.[64,65] SPECT typically shows reduced perfusion of the affected brain interictally and increased perfusion during a seizure.[66,67] Functional imaging may not be necessary for routine patient

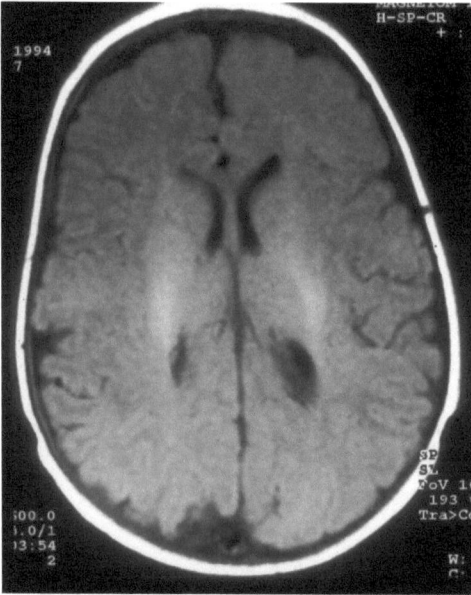

Figure 4-2. A: Magnetic resonance study from a child with Sturge-Weber syndrome; T1-weighted axial view without contrast infusion.

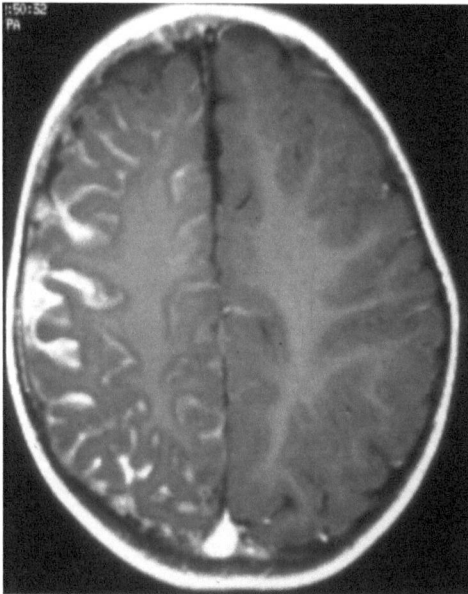

Figure 4-2. B: On the axial view with gadolinium, his scan reveals a more obvious leptomeningeal angioma.

care, but may be a helpful tool during the evaluation before epilepsy surgery.[67]

Electroencephalography (EEG) is usually abnormal in people with Sturge-Weber syndrome, but the EEG findings are similar to the abnormalities recorded in other patients with epileptic seizures and the test is sometimes

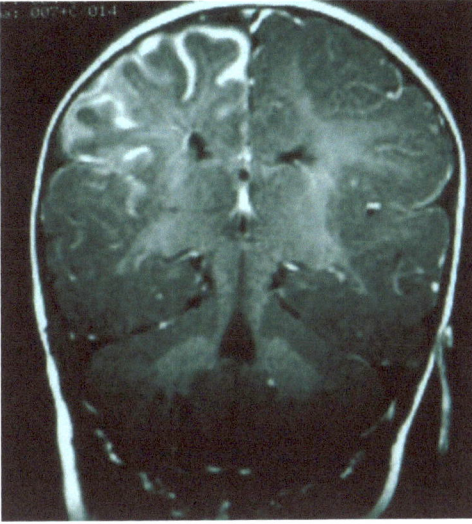

Figure 4-2. C: Coronal section with gadolinium of another patient demonstrating the usefulness of this section in delineating the extent of the lesion.

normal. Nevertheless, EEG can help to establish the diagnosis in children with subtle findings. The amplitude adjacent to the leptomeningeal angioma is often reduced.[20,68] Focal epileptiform discharges may be found either ipsilateral[68] or contralateral[20] to the most severely affected hemisphere. Synchronous bilateral discharges are sometimes recorded even in patients with radiographic involvement of only one hemisphere.[19] The EEG pattern sometimes makes it easier to select the most effective anticonvulsant medication, and it is an essential tool when trying to localize the site of onset and extent of seizure activity prior to surgery.

The cerebrospinal fluid (CSF) protein level may be elevated, although this is a nonspecific abnormality without much practical use.[69,70] The CSF pressure may be elevated in the few patients who develop hydrocephalus. Otherwise, the CSF is typically normal.

Presymptomatic Diagnosis

Early presymptomatic diagnosis of Sturge-Weber syndrome brain involvement (and alternatively excluding brain involvement) in infants with facial port-wine birthmarks remains an issue despite the wide availability of CT and MRI because these studies are frequently normal in at-risk newborns and young infants and it is unclear at what point a normal contrast-enhanced MRI of the brain absolutely excludes Sturge-Weber syndrome brain involvement. Based on our combined clinical experience and the literature, an at-risk child with a facial port-wine birthmark with normal development, no history of seizures, and a

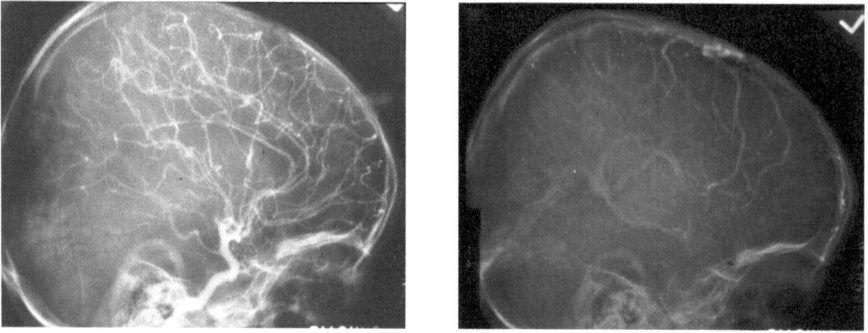

Figure 4-3. A: The arterial phase of this Sturge-Weber patient's angiogram is normal. **B:** During the venous phase, however, cortical veins are not seen in the posterior hemisphere. *Reprinted from* Childs Brain *with permission of Karger Basil*

normal contrast enhanced MRI of the brain after a year of age probably does not have brain involvement.

Currently under development, quantitative EEG is a promising screening approach to enhance our ability to provide earlier reassurance and select infants for imaging prior to a year of age. One report applied quantitative EEG to a cohort of nine infants with facial port-wine nevi in an effort to identify the individuals with brain lesions.[71] If this technique could be applied to larger numbers of infants, asymmetries of EEG or power calculations on quantitative EEG could become a useful tool for screening at-risk newborns and young infants to determine who should undergo early MRI imaging and which patients can wait for later confirmatory imaging.

NEUROPATHOLOGY OF STURGE-WEBER SYNDROME

Although any part of the brain can be affected, the occipital and parietal lobes are more often involved than the frontal lobes.[72] Diffuse involvement of one or both cerebral hemispheres occurs less often. Angiomatous vessels of the leptomeninges may obliterate the subarachnoid space, and the tortuous deep-draining veins that are seen radiographically can also be seen in pathologic specimens.[73-75]

Grossly, the leptomeninges appear thickened and discolored because of the increased vascularity of the tissue (**Figure 4-4**). The underlying parenchyma may be atrophic and contain multiple calcific granular deposits (**Figures 4-5 & 4-6**).[73,74,76] Cerebral atrophy may become progressively more severe during early childhood before eventually stabilizing, and children with mild clinical features

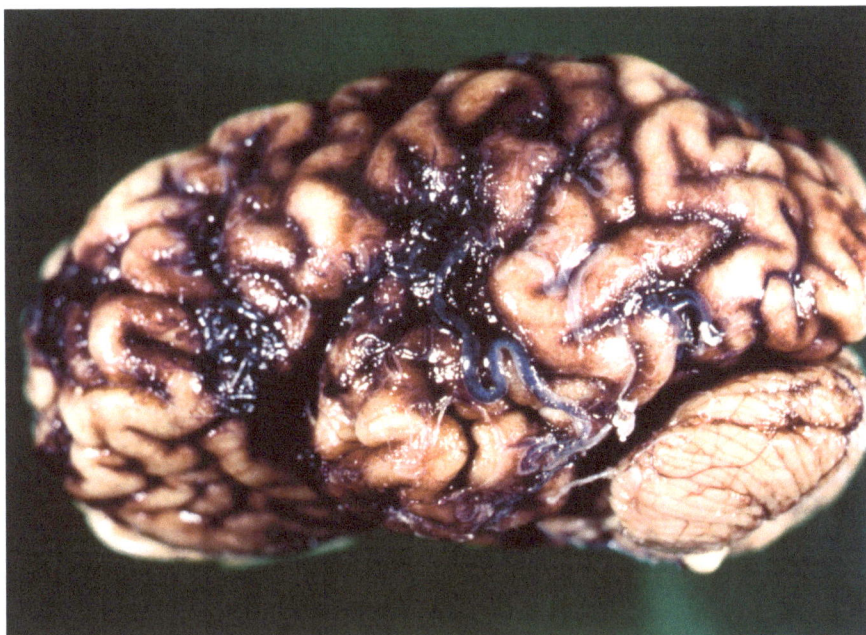

Figure 4-4. The left lateral surface of the brain of a patient with Sturge-Weber syndrome. Note the thickened discolored leptomeninges and the large tortuous venous structures visible over the surface of the lesion.
S. S. Schochet

may not develop visible atrophy at all. Microscopic examination reveals neuronal loss and gliosis, which, like the angiomatosis, usually extends beyond the area of radiographic abnormality. The typical pattern calcification within the gyrus results from deposition of calcium within the outer cerebral cortical layers.[73,76]

In addition to the vascular abnormalities, dysplasia of the cerebral tissue is common. Maton and colleagues identified cortical dysplasia and/ or polymicrogyria in all six of the available brain specimens from a cohort of Sturge-Weber patients undergoing epilepsy surgery.[27] The occurrence of cerebral dysplasia in these individuals provides another possible avenue for intractable epilepsy and suggests that therapy directed toward the vascular pathology alone is unlikely to interrupt intractable epilepsy.

Microscopically these vessels are primarily thin-walled veins of variable size.[73,76] Angiomatous vessels sometimes extend into the superficial brain parenchyma, often extending into the normal-looking areas adjacent to the visible malformation.[75] The ipsilateral choroid plexus is frequently affected. Some vessels are narrowed or occluded by progressive hyalinization and subendothelial proliferation,[73,77] although there is seldom pathologic evidence of recent cerebral infarction.

Recent molecular neuropathology studies suggest that the leptomeningeal angioma of Sturge-Weber syndrome undergoes angiogenic remodeling rather than being a static vascular malformation.[78,79] Comati and colleagues documented increased expression of VEGF and its receptors as well as the angiopoitin receptor Tie2 in the Sturge-Weber syndrome leptomeningeal vessels.[78] These findings were associated with elevated HIF-1α and HIF-2α expression in the endothelial cells of Sturge-Weber syndrome leptomeningeal vessels and both increased an proliferative index and evidence of apoptosis in these vessels. Together, these findings suggested a model where active vascular remodeling, perhaps with VEGF signaling at its center, contributes to the maintenance and progression of the vascular malformation and neurologic morbidity. However, it is too early to advocate either anti-VEGF or anti-HIF therapy; this vascular remodeling could serve a partly compensatory role, so that such treatment would have entirely unexpected and adverse results.

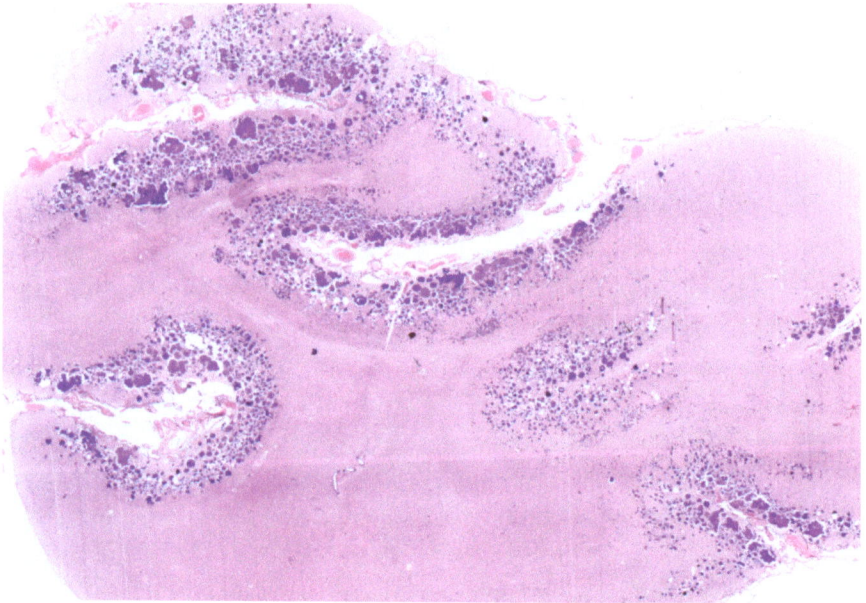

Figure 4-5. Section of cerebrum from a patient with Sturge-Weber syndrome. Note the leptomeningeal angioma filling the sulci and confined to the subarachnoid space. The dark staining granular material in the underlying cortex represents dystrophic calcification of the parenchymal tissue.
S. S. Schochet

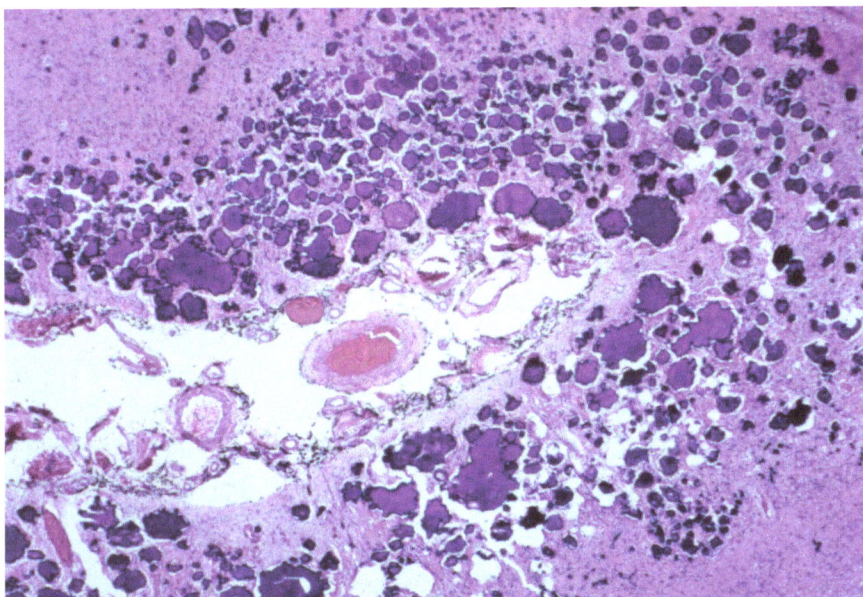

Figure 4-6. High magnification of a leptomeningeal angioma from a patient with Sturge-Weber syndrome. The angioma is largely limited to the subarachnoid space and the dystrophic calcifications to the underlying parenchymal tissue. *S. S. Schochet*

TREATMENT OF NEUROLOGIC COMPLICATIONS

Given that chronically impaired perfusion punctuated by intermittent ischemia is thought to have a role in the neurologic deterioration suffered by some young children with Sturge-Weber syndrome, and seizures and decompensation often occur in the setting of illness, general recommendations for stroke prevention and maintenance of health should apply. Thus, iron deficiency anemia should be treated during the first year of life, hydration must be maintained aggressively, fevers should be treated during illnesses. If excessive snoring is present, a sleep study should be considered to exclude obstructive sleep apnea. Children with Sturge-Weber syndrome should generally receive the influenza vaccine and other standard immunizations.

Anticonvulsant Medications

Effective seizure control can limit neurologic impairment and improve the quality of life for these patients. As a rule children with a more extensive intracranial lesion tend to have more difficult-to-control seizures. Several new antiepileptic medications have been released recently, and complete seizure control may not be as difficult as was once suggested. In one series half of the patients achieved complete seizure control and an additional 39% had partial control.[3]

Careful attention to weight-based dosing of antiepileptic agents and optimization of dosing schedules assures the best possible seizure control. Parents of at-risk infants should be taught to recognize seizures and the need to quickly seek medical attention for them. Some physicians provide the family with a rescue medication (e.g. rectal diazepam or intranasal midazolam) in case the first seizure is prolonged, although the evidence supporting this approach is sparse.

Ville and colleagues compared 16 patients who were treated with barbiturates before the onset of seizures to 21 individuals who were already taking an anticonvulsant at the time of presentation. This intriguing study suggested that the pretreated subjects were less likely to become mentally retarded, but a number of limitations make it difficult to confidently advocate the use of medication before the onset of seizures.[80] The age when the children began treatment is unclear. The study was not randomized and the pretreated group had more subjects with more limited intracranial involvement, creating the potential for a better cognitive outcome in this group. Nevertheless, presymptomatic diagnosis and neuroprotective treatment may be an important topic for future study.

Surgical Treatment

Lesion resection usually improves seizure control and may enhance intellectual development.[81,82] The operative risk from hemispherectomy includes intracranial hemorrhage and infection, but in experienced hands these risks are low enough to outweigh the potential benefits of the surgery for an individual with medically intractable epilepsy. Resection of the most severely affected area may achieve acceptable results with fewer surgical complications and with fewer new neurological deficits, although recent studies show clearly that long-term seizure freedom is clearly associated with complete resection of the involved brain regions.[25,83-85]

Most physicians do not feel comfortable recommending surgery for a patient who has not yet developed seizures, or one whose seizures are well controlled with medication There is also understandable reluctance to resect a still-functional portion of the brain and cause a new deficit.[86] Thus surgery is often reserved for patients with severe seizures who have not responded to medication and who already have dysfunction of the area to be removed (e.g. hemiparesis or hemianopia).

Surgical resection may not be appropriate for children with extensive bilateral brain involvement, but resection of a severely affected region of one hemisphere is sometimes beneficial even in an individual with bilateral disease.[87] Corpus callosotomy is an alternative for some individuals with extensive bilateral disease, or those with refractory tonic or atonic seizures in whom more definitive surgery is not feasible.[88] Some patients who are not appropriate for lesion resection may benefit from a vagus nerve stimulator.

Despite the general agreement that surgical resection is effective, there is still debate about patient selection and about the timing of surgery.[89,90] Surgical guidelines have been developed for Sturge-Weber patients. Hemispherectomy should be considered for patients with clinically significant seizures who fail to respond to an adequate trial of anticonvulsants.[89] Some physicians believe that daily aspirin improves seizure control in a portion of the patients. There is some evidence that early hemispherectomy allows better cognitive development of children whose seizures begin in infancy,[91,92] but surgery may not be necessary in children whose seizures are fully controlled with medication. Surgery should only be done in a center with an ongoing program in pediatric epilepsy surgery and age-appropriate facilities for preoperative and postoperative care. Patients with a very localized lesion should have a limited resection, rather than a complete hemispherectomy, preserving as much normal brain as possible, even at the risk of having to do additional surgery later. Corpus callosotomy should be reserved for patients with refractory tonic or atonic seizures. In effect, a similar approach should be used in children with Sturge-Weber syndrome as with other epileptic patients.[89] Surgical treatment of epilepsy caused by Sturge-Weber syndrome is presented in more detail in Chapter 7.

Headache Treatment

Headaches and migraines in Sturge-Weber syndrome are treated with standard abortive medications and, when necessary, preventative medications. One survey of 74 individuals with Sturge-Weber syndrome and migraine noted that 22% were treated with triptan agents and that the majority reported relief without complications.[93] Although the prevalence of weakness and worsening of stroke-like events is low, the response to triptans in patients with Sturge-Weber syndrome should be closely monitored. It is not certain that the pathogenesis of headache is the same in individuals with Sturge-Weber syndrome, so these agents may need to be studied further.

Slightly more than a third of the individuals in this survey reported using a headache preventative medication. Medications such as topiramate, valproate and gabapentin are often used to prevent both the seizures and the headaches. In some patients, headache can precede a stroke-like episode or seizure, and particularly in these patients headaches should be aggressively treated and prevented if possible.

Antiplatelet Agents

Daily aspirin has been used empirically for several years to prevent recurrent venous thromboses that are suspected to cause neurologic deterioration.[30,94,95] Some children taking daily aspirin have done well, but the complete lack of controlled clinical trials and the well documented clinical variability of the syndrome make it very difficult to determine whether aspirin helps children with Sturge-Weber syndrome. Nevertheless, it seems reasonable to use aspirin, at least in patients with repeated clinical episodes of transient neurologic deficits, and perhaps for patients with bihemispheric disease for whom surgery is not a reasonable option. Most children tolerate low-dose daily aspirin well, although the optimum dose has not been clearly established. One small open-label study found a reduction in stroke-like episodes in patients treated with low-dose aspirin.[96] Another small series of 6 subjects attributed decreased seizures to the use of low-dose aspirin.[97] Larger prospective open-label studies are currently underway, but randomized, placebo-controlled studies will likely be needed to fully establish the utility of low-dose aspirin and identify which individuals are most likely to benefit from it. The newer antiplatelet drugs have yet to be studied and cannot be endorsed at this time.

Neuroendocrine Concerns

An increased prevalence of growth hormone deficiency (18-fold higher than the general population) has been documented in Sturge-Weber syndrome.[98] Neuroimaging did not show obvious abnormalities of the pituitary gland or hypothalamus, so the reason for the deficiency remains unclear. Nevertheless, growth hormone deficiency should be considered in individuals with short stature or growth failure.

Central hypothyroidism has been reported in individuals with Sturge-Weber syndrome, although the reason for it is uncertain.[99] All of the children in whom this has been diagnosed have been treated with chronic anticonvulsants, and central hypothyroidism has been associated with the use of several of the anticonvulsants. Given the increased risk of growth hormone deficiency in Sturge-Weber syndrome, disruption of the hypothalamic-pituitary axis is also possible. Central hypothyroidism in the setting of Sturge-Weber syndrome can be easy to overlook because some of its symptoms and signs (e.g., inattention, sleepiness, obesity, headaches, and behavioral issues) can result from anticonvulsants, seizures, Sturge-Weber syndrome-related brain dysfunction, and behavioral problems.

REFERENCES

1. Alexander GL, Norman RM. *Sturge-Weber Syndrome*. Bristol: John Wright & Sons Ltd, 1960.

2. Roach ES. Neurocutaneous syndromes. *Pediatr Clin North Am* 1992;39:591-620.

3. Sujansky E, Conradi S. Sturge-Weber syndrome: age of onset of seizures and glaucoma and the prognosis for affected children. *J Child Neurol* 1995;10:49-58.

4. Sujansky E, Conradi S. Outcome of Sturge-Weber syndrome in 52 adults. *Am J Med Genetics* 1995;57:35-45.

5. Peterman AF, Hayles AB, Dockerty MB, Love JG. Encephalotrigeminal angiomatosis (Sturge-Weber disease). *J Am Med Assoc* 1958;167:2169-2176.

6. Uram M, Zubillaga C. The cutaneous manifestations of Sturge-Weber syndrome. *J Clin Neuro Ophthalmol* 1982;2:245-248.

7. Enjolras O, Riche MC, Merland JJ. Facial port-wine stains and Sturge-Weber syndrome. *Pediatrics* 1985;76:48-51.

8. Tallman B, Tan OT, Morelli JG et al. Location of port-wine stains and the likelihood of ophthalmic and/or central nervous system complications. *Pediatrics* 1991;87:323-327.

9. Crosley CJ, Binet EF. Sturge-Weber Syndrome- presentation as a focal seizure disorder without nevus flammeus. *Clin Pediatr* 1978;17:606-609.

10. Gorman RJ, Snead OC. Sturge-Weber syndrome without port-wine nevus (letter). *Pediatrics* 1977;60:785.

11. Taly AB, Nagaraja D, Das S, Shankar SK, Pratibha NG. Sturge-Weber-Dimitri disease without facial nevus. *Neurology* 1987;37:1063-1064.

12. Simmat G, Lelong B, Morin M. Peculiar clinical and x-ray computed tomographic aspects in Sturge-Weber disease. Bilateral occipital calcifications without facial angioma. *J Radiolog* 1984;65:279-283.

13. Boltshauser E, Wilson J, Hoare RD. Sturge-Weber syndrome with bilateral intracranial calcification. *J Neurol Neurosurg Psychiatry* 1976;39:429-435.

14. Cersoli M, Campanile S, Campanile A, Amore M. Unusual findings in Sturge-Weber syndrome. *AJNR Am J Neuroradiol* 1989;10:S85.

15. Bebin EM, Gomez MR. Prognosis in Sturge-Weber disease: comparison of unihemispheric and bihemispheric involvement. *J Child Neurol* 1988;3:181-184.

16. Oakes WJ. The natural history of patients with the Sturge-Weber syndrome. *Pediatr Neurosurg* 1992;18:287-290.

17. Kramer U, Kahana E, Shorer Z, Ben-Zeev B. Outcome of infants with unilateral Sturge-Weber syndrome and early onset seizures. *Dev Med Child Neurol* 2000;42:756-759.

18. Welch K, Naheedy MH, Abroms IF, Strand RD. Computed tomography of Sturge-Weber syndrome. *J Comput Assist Tomogr* 1980; 4:33-36.

19. Chevrie JJ, Specola N, Aicardi J. Secondary bilateral synchrony in unilateral pial angiomatosis: Successful surgical management. *J Neurol Neurosurg Psychiatry* 1988;15:95-98.

20. Fukuyama Y, Tsuchiya S. A study on Sturge-Weber syndrome. *Eur Neurol* 1979;18:194-204.

21. Miyama S, Goto T. Leptomeningeal angiomatosis with infantile spasms. *Pediatr Neurol* 2004;31:353-356.

22. Barbagallo M, Ruggieri M, Incorpora G et al. Infantile spasms in the setting of Sturge-Weber syndrome. *Childs Nerv Syst* 2009;25:111-118.

23. Ewen JB, Comi AM, Kossoff EH. Myoclonic-astatic epilepsy in a child with Sturge-Weber syndrome. *Pediatr Neurol* 2007;36:115-117.

24. Kossoff EH, Ferenc L, Comi AM. An infantile-onset, severe, yet sporadic seizure pattern is common in Sturge-Weber syndrome. *Epilepsia* 2009;50:2154-2157.

25. Aicardi J, Arzimanoglou A. Sturge-Weber syndrome. *International Pediatrics* 1991;6:129-134.

26. Chapieski L, Friedman A, Lachar D. Psychological functioning in children and adolescents with Sturge-Weber syndrome. *J Child Neurol* 2000;15:660-665.

27. Maton B, Krsek P, Jayakar P et al. Medically intractable epilepsy in Sturge-Weber syndrome is associated with cortical malformation: implications for surgical therapy. *Epilepsia* 2010;51:257-267.

28. Lee S. Psychopathology in Sturge-Weber syndrome. *Can J Psychiat* 1990;35:674-678.

29. Jansen FE, van der Worp HB, van HA, van NO. Sturge-Weber syndrome and paroxysmal hemiparesis: epilepsy or ischaemia? *Dev Med Child Neurol* 2004;46:783-786.

30. Garcia JC, Roach ES, McLean WT. Recurrent thrombotic deterioration in the Sturge-Weber syndrome. *Childs Brain* 1981;8:427-433.

31. Zolkipli Z, Aylett S, Rankin PM, Neville BG. Transient exacerbation of hemiplegia following minor head trauma in Sturge-Weber syndrome. *Dev Med Child Neurol* 2007;49:697-699.

32. Fishman MA, Baram TZ. Megalencephaly due to impaired cerebral venous return in a Sturge-Weber variant syndrome. *J Child Neurol* 1986; 1:115-118.

33. Meyer E. Neurocutaneous syndrome with excessive macrohydrocephalus (Sturge-Weber/Klippel-Trenaunay syndrome). *Neuropadiatrie* 1979;10:67-75.

34. Shapiro K, Shulman K. Facial nevi associated with anomalous venous return and hydrocephalus. *J Neurosurg* 1976;45:20-25.

35. Stephan MJ, Hall BD, Smith DW, Cohen MM. Macrocephaly in association with unusual cutaneous angiomatosis. *J Pediatr* 1975;87:353-359.

36. Orr LS, Osher RH, Savino PJ. The syndrome of facial nevi, anomalous venous return and hydrocephalus. *Ann Neurol* 1978;3:316-318.

37. Klapper J. Headache in Sturge-Weber syndrome. *Headache* 1994;34:521-522.

38. Kossoff EH, Hatfield LA, Ball KL, Comi AM. Comorbidity of epilepsy and headache in patients with Sturge-Weber syndrome. *J Child Neurol* 2005;20:678-682.

39. Riela AR, Stump D, Roach ES, McLean WT, Garcia JC. Regional cerebral blood flow characteristics of the Sturge-Weber syndrome. *Pediatr Neurol* 1985;1:85-90.

40. Okudaira Y, Arai H, Sato K. Hemodynamic compromise as a factor in clinical progression of Sturge-Weber syndrome. *Child's Nerv Syst* 1997;13:214-219.

41. Iizuka T, Sakai F, Yamakawa K, Suzuki K, Suzuki N. Vasogenic leakage and the mechanism of migraine with prolonged aura in Sturge-Weber syndrome. *Cephalalgia* 2004;24:767-770.

42. Cushing H. Cases of spontaneous intracranial hemorrhage associated with trigeminal nevi. *JAMA* 1906;47:178-183.

43. Di Chiro G, Lindgren E. Radiographic findings in 14 cases of Sturge-Weber syndrome. *Acta Radiol* 1951;35:387-399.

44. Anderson FH, Duncan GW. Sturge-Weber disease with subarachnoid hemorrhage. *Stroke* 1974;5:509-511.

45. Juhász C, Batista CE, Chugani DC, Muzik O, Chugani HT. Evolution

of cortical metabolic abnormalities and their clinical correlates in Sturge-Weber syndrome. *Eur J Paediatr Neurol* 2007;11:277-284.

46. Aylett SE, Neville BG, Cross JH, Boyd S, Chong WK, Kirkham FJ. Sturge-Weber syndrome: cerebral haemodynamics during seizure activity. *Dev Med Child Neurol* 1999;41:480-485.

47. Namer IJ, Battaglia F, Hirsch E, Constantinesco A, Marescaux C. Subtraction ictal SPECT co-registered to MRI (SISCOM) in Sturge-Weber syndrome. *Clin Nucl Med* 2005;30:39-40.

48. Chiron C, Raynaud C, Dulac O, Tzourio N, Plouin P, Tran-Dinh S. Study of the cerebral blood flow in partial epilepsy of childhood using the SPECT method. *J Neuroradiol* 1989;16:317-324.

49. Hu J, Yu Y, Juhász C et al. MR susceptibility weighted imaging (SWI) complements conventional contrast enhanced T1 weighted MRI in characterizing brain abnormalities of Sturge-Weber Syndrome. *J Magn Reson Imaging* 2008;28:300-307.

50. Marti-Bonmati L, Menor F, Poyatos C, Cortina H. Diagnosis of Sturge-Weber syndrome: comparison of the efficacy of CT and MR imaging in 14 cases. *Am J Roentgenol* 1993;158:867-871.

51. Wasenko JJ, Rosenbloom SA, Duchesneau PM, Lanzieri CF, Weinstein MA. The Sturge-Weber syndrome: Comparison of MR and CT characteristics. *AJNR Am J Neuroradiol* 1990;11:131-134.

52. Chamberlain MC, Press GA, Hesselink JR. MR imaging and CT in three cases of Sturge-Weber syndrome: prospective comparison. *Am J Roentgenol* 1989;10:491-496.

53. Lipski S, Brunelle F, Aicardi J, Hirsch JF, Lallemand D. Gd-DOTA-enhanced MR imaging in two cases of Sturge- Weber syndrome. *AJNR Am J Neuroradiol* 1990;11:690-692.

54. Elster AD, Chen MY. MR imaging of Sturge-Weber syndrome: role of gadopentetate dimeglumine and gradient-echo techniques. *AJNR Am J Neuroradiol* 1990;11:685-689.

55. Benedikt RA, Brown DC, Walker R, Ghaed VN, Mitchell M, Geyer CA. Sturge-Weber syndrome: cranial MR imaging with Gd-DTPA. *AJNR Am J Neuroradiol* 1993;14:409-415.

56. Sperner J, Schmauser I, Bittner R et al. MR-imaging findings in children with Sturge-Weber syndrome. *Neuropediatrics* 1990;21:146-152.

57. Griffiths PD, Coley SC, Romanowski CA, Hodgson T, Wilkinson ID. Contrast-enhanced fluid-attenuated inversion recovery imaging for leptomeningeal disease in children. *AJNR Am J Neuroradiol* 2003;24:719-723.

58. Mentzel HJ, Dieckmann A, Fitzek C, Brandl U, Reichenbach JR, Kaiser WA. Early diagnosis of cerebral involvement in Sturge-Weber syndrome using high-resolution BOLD MR venography. *Pediatr Radiol* 2005;35:85-90.

59. Terdjman P, Aicardi J, Sainte-Rose C, Brunelle F. Neuroradiological findings in Sturge-Weber syndrome (SWS) and isolated pial angiomatosis. *Neuropediatrics* 1991;22:115-120.

60. Probst FP. Vascular morphology and angiographic flow patterns in Sturge-Weber angiomatosis. *Neurorad* 1980;20:73-78.

61. Bentson JR, Wilson GH, Newton TH. Cerebral venous drainage pattern of the Sturge-Weber syndrome. *Radiology* 1971;101:111-118.

62. Poser CM, Taveras JM. Cerebral angiography in encephalo-trigeminal angiomatosis. *Radiology* 1957;68:327-336.

63. Hamano K, Ito M, Inai K, Nose T, Takita H. A case of Sturge-Weber syndrome with peculiar venous abnormalities. *Child Nerv Sys* 1993;9:491-493.

64. Chugani HT, Mazziotta JC, Phelps ME. Sturge-Weber syndrome: a study of cerebral glucose utilization with positron emission tomography. *J Pediatr* 1989;114:244-253.

65. Reid DE, Maria BL, Drane WE, Quisling RG, Hoang KB. Central nervous system perfusion and metabolism abnormalities in Sturge-Weber syndrome. *J Child Neurol* 1997;12:218-222.

66. Chiron C, Raynaud C, Tzourio N et al. Regional cerebral blood flow by SPECT imaging in Sturge-Weber disease: an aid for diagnosis. *J Neurol Neurosurg Psychiatry* 1989;52:1402-1409.

67. Bilgin O, Vollmar C, Peraud A, la FC, Beleza P, Noachtar S. Ictal SPECT in Sturge-Weber syndrome. *Epilepsy Res* 2008;78:240-243.

68. Brenner RP, Sharbrough FW. Electroencephalographic evaluation in Sturge-Weber syndrome. *Neurology* 1976;26:629-632.

69. Skoglund RR, Paa D, Lewis WJ. Elevated spinal-fluid protein in Sturge-Weber syndrome. *Develop Med Child Neurol* 1978;20:99-102.

70. Boltshauser E, Wilson J. Elevated spinal-fluid protein in Sturge-Weber syndrome. *Develop Med Child Neurol* 1978;20:392-393.

71. Ewen JB, Kossoff EH, Crone NE et al. Use of quantitative EEG in infants with port-wine birthmark to assess for Sturge-Weber brain involvement. *Clin Neurophysiol* 2009;120:1433-1440.

72. Hatfield M, Muraki A, Wollman R, Hekmatpanah J, Mojtahedi S, Duda EE. Isolated frontal lobe calcification in Sturge-Weber syndrome. *AJNR Am J Neuroradiol* 1988;9:203-204.

73. Wohlwill FJ, Yakovlev PI. Histopathology of meningo-facial

angiomatosis (Sturge-Weber's disease). *J Neuropathol Exp Neurol* 1957;16:341-364.

74. Nellhaus G, Haberland C, Hill BJ. Sturge-Weber disease with bilateral intracranial calcifications at birth and unusual pathologic findings. *Acta Neurol Scand* 1967;43:314-347.

75. Roizin L, Gold G, Herman HH, Bonafede VI. Congenital vascular anomalies and their histopathology in Sturge- Weber-Dimitri syndrome. *J Neuropathol Exp Neurol* 1959;18:75-97.

76. Di Trapani G, Di Rocco C, Abbamondi AL, Caldarelli M, Pocchiari M. Light microscopy and ultrastructural studies of Sturge-Weber disease. *Brain* 1982;9:23-36.

77. Norman MG, Schoene WC. The ultrastructure of Sturge-Weber disease. *Acta Neuropathol* 1977;37:199-205.

78. Comati A, Beck H, Halliday W, Snipes GJ, Plate KH, Acker T. Upregulation of hypoxia-inducible factor (HIF)-1alpha and HIF-2alpha in leptomeningeal vascular malformations of Sturge-Weber syndrome. *J Neuropathol Exp Neurol* 2007;66:86-97.

79. Comi AM, Weisz CJ, Highet BH, Skolasky RL, Pardo CA, Hess EJ. Sturge-Weber syndrome: altered blood vessel fibronectin expression and morphology. *J Child Neurol* 2005;20:572-577.

80. Ville D, Enjolras O, Chiron C, Dulac O. Prophylactic antiepileptic treatment in Sturge-Weber disease. *Seizure* 2002;11:145-150.

81. Falconer MA, Rushworth RG. Treatment of encephalotrigeminal angiomatosis (Sturge-Weber disease) by hemispherectomy. *Arch Dis Child* 1960;35:433-447.

82. Ogunmekan AO, Hwang PA, Hoffman HJ. Sturge-Weber-Dimitri disease: Role of hemispherectomy in prognosis. *Can J Neurol Sci* 1989;16:78-80.

83. Rosen I, Salford L, Starck L. Sturge-Weber disease-neurophysiological evaluation of a case with secondary epileptogenesis, successfully treated with lobe-ectomy. *Neuropediatrics* 1984;15:95-98.

84. Bye AM, Matheson JM, Mackenzie RA. Epilepsy surgery in Sturge-Weber syndrome. *Austral N Zealand J Ophthal* 1989;25:103-105.

85. Schropp C, Sorensen N, Krauss J. Early periinsular hemispherotomy in children with Sturge-Weber syndrome and intractable epilepsy--outcome in eight patients. *Neuropediatrics* 2006;37:26-31.

86. Arzimanoglou A, Aicardi J. The epilepsy of Sturge-Weber syndrome: Clinical features and treatment in 23 patients. *Acta Neurol Scand (Suppl)* 1992;140:18-22.

87. Hallbook T, Ruggieri P, Adina C et al. Contralateral MRI abnormalities in candidates for hemispherectomy for refractory epilepsy. *Epilepsia* 2010;51:556-563.

88. Rappaport ZH. Corpus callosum section in the treatment of intractable seizures in the Sturge-Weber syndrome. *Child Nerv Sys* 1988;4:231-232.

89. Roach ES, Riela AR, Chugani HT, Shinnar S, Bodensteiner JB, Freeman J. Sturge-Weber syndrome: recommendations for surgery. *J Child Neurol* 1994;9:190-193.

90. Kossoff EH, Buck C, Freeman JM. Outcomes of 32 hemispherectomies for Sturge-Weber syndrome worldwide. *Neurology* 2002;59:1735-1738.

91. Arzimanoglou AA, Andermann F, Aicardi J et al. Sturge-Weber syndrome: indications and results of surgery in 20 patients. *Neurology* 2000;55:1472-1479.

92. Steinbok P, Gan PY, Connolly MB et al. Epilepsy surgery in the first 3 years of life: a Canadian survey. *Epilepsia* 2009;50:1442-1449.

93. Kossoff EH, Balasta M, Hatfield LM, Lehmann CU, Comi AM. Self-reported treatment patterns in patients with Sturge-Weber syndrome and migraines. *J Child Neurol* 2007;22:720-726.

94. McCaughan RA, Ouvrier RA, De Silva K, McLaughlin A. The value of the brain scan and cerebral arteriogram in the Sturge-Weber syndrome. *Proc Aust Assoc Neurol* 1975;12:185-190.

95. Roach ES, Riela AR, McLean WT, Stump DA. Aspirin therapy for Sturge-Weber syndrome (abstract). *Ann Neurol* 1985;18:387.

96. Maria BL, Neufeld JA, Rosainz LC et al. Central nervous system structure and function in Sturge-Weber syndrome: evidence of neurologic and radiographic progression. *J Child Neurol* 1998;13:606-618.

97. Udani V, Pujar S, Munot P, Maheshwari S, Mehta N. Natural history and magnetic resonance imaging follow-up in 9 Sturge-Weber Syndrome patients and clinical correlation. *J Child Neurol* 2007;22:479-483.

98. Miller RS, Ball KL, Comi AM, Germain-Lee EL. Growth hormone deficiency in Sturge-Weber syndrome. *Arch Dis Child* 2006;91:340-341.

99. Comi AM, Bellamkonda S, Ferenc LM, Cohen BA, Germain-Lee EL. Central hypothyroidism and Sturge-Weber syndrome. *Pediatr Neurol* 2008;39:58-62.

Chapter 5

Cognitive and Behavioral Aspects of Sturge-Weber Syndrome

Lynn Chapieski, Ph.D.

As with most other neurological disorders, the characteristic features of Sturge-Weber syndrome extend beyond the purely medical. Individuals with Sturge-Weber syndrome are at risk for academic, behavior, and social problems and, in many cases, these associated disorders present the greatest challenges to the patients and their families. Factors that increase the risk of academic and social problems potentially include extensive neurological involvement, seizures, medication side effects, cosmetic issues, and the stresses of living with a chronic medical condition.

Because of the rarity of Sturge-Weber syndrome, data on its behavioral effects are limited. Current knowledge comes from a small number of studies of Sturge-Weber syndrome and what can be inferred from studies of individuals who do not have the syndrome but have seizure disorders, port-wine stains, or other related medical conditions. This chapter will review the evidence for academic, cognitive, mood, behavior, and social problems in individuals with Sturge-Weber syndrome as well as the associated risk factors.

Academic and Cognitive Functioning

Developmental and school problems are common in students with Sturge-Weber syndrome. The first large-scale reports of neurological and behavioral characteristics of individuals with Sturge-Weber syndrome were conducted by Sujansky and Conradi.[1,2] These investigators collected survey information from about 171 individuals ranging in age from 2 months to 59 years of age and then followed up with telephone interviews of those who were at least 18 years of age. The telephone interviews were conducted with either the affected adult or a caretaker. Fifty-eight percent of the larger sample had experienced delayed early developmental milestones and received special education services. Only 65% of the adults had received a high school diploma, although 26% had enrolled in either a two-year or a four-year college.

In a more recent survey of the parents and teachers of 79 school-aged children with Sturge-Weber syndrome, 65% of parents reported that their children were receiving special education services and 70% reported that they were at least moderately concerned about school problems.[3] Both parents and teachers reported significantly more academic problems for the affected child than for unaffected siblings.

The frequency of academic problems is not surprising given the high incidence of mental retardation. Studies have consistently estimated the incidence of mental retardation in individuals with Sturge-Weber syndrome to be 50% to 60%.[1-5] Even though there is a high incidence of mental retardation, the range of intellectual functioning is wide, and some individuals with Sturge-Weber syndrome have intellectual functioning in the superior range.[3] Intellectual functioning is highly correlated with school performance in children with Sturge-Weber syndrome just as it is in the general population.[3]

A diagnosis of mental retardation indicates a global impairment of cognitive abilities, but even more circumscribed cognitive impairments can interfere with the acquisition of one or more academic skills. Because individuals with epilepsy and many other neurological conditions have an increased incidence of specific cognitive problems and learning disabilities,[6,7] it is likely that that individuals with Sturge-Weber syndrome are also at increased risk. This may apply to even those individuals with Sturge-Weber syndrome who have higher levels of intellectual functioning. However, little is known about the incidence or types of specific learning disabilities associated with Sturge-Weber syndrome.

Although detailed information about specific cognitive impairments associated with Sturge-Weber syndrome is lacking, it is unlikely that there are educational needs that are entirely unique to this group. Parents and school personnel sometimes confuse the neurological disorder with the functional disability. Educators who are understandably nervous about a rare neurological diagnosis may overlook the more commonplace educational handicap of mental retardation. Parents sometimes report that school personnel do not know how to meet the educational needs of their child because of their lack of familiarity with Sturge-Weber syndrome. In this instance, medical professionals should help parents and teachers to focus on the functional disability rather than the neurological diagnosis.

Risk Factors for Cognitive Impairment

Data on risk factors for cognitive impairment are limited but the presence of epilepsy appears to increase the risk for developmental and academic problems. Sujansky and Conradi[1] reported early developmental delays in 71% of their sample who had epilepsy and in only 6% of those who had not developed a seizure disorder. Only 11% of their sample without epilepsy required special education, while 69% of those with epilepsy had special education needs. None of the patients described by Bebin and Gomez who had not developed seizures had subnormal intelligence.[5] Chapieski and colleagues also identified the presence of a seizure disorder as an important prognostic indicator for cognitive and academic development, whether assessed by parent and teacher responses to a questionnaire or by direct assessment of intelligence.[3]

Raches found no significant difference between parent and teacher reports of intellectual and academic performance for children with Sturge-Weber syndrome without a seizure disorder and a demographically similar group of normal children, suggesting that individuals with Sturge-Weber syndrome but not epilepsy have a similar incidence of such problems as members of the general population.[8] **Table 5-1** illustrates average parent and teachers reports of cognitive and academic problems for the two groups.

Parallels between cognitive ability and seizure-related variables, such as seizure frequency, age of onset, and duration of epilepsy, have been less predictable. A few studies have linked an earlier age of seizure onset to poor developmental outcome.[1,9,11] Other studies have failed to confirm a relationship between age of seizure onset and cognitive ability, although studies with negative findings tend to have had smaller sample sizes.[10,12]

An association between seizure control or seizure frequency and cognition has not generally been supported, with one exception.[3,10,12] In a cohort of 55 patients, Pascual-Castroviejo and colleagues noted lower mental development in individuals with refractory seizures.[9] Although Kramer and colleagues did not find that ongoing seizures *per se* were related to cognitive problems, indicators of stronger seizure intensity (frequency, duration of single seizures, and duration of bouts of seizures, presence of secondary generalization, and occurrence of status epilepticus) at early ages did have a deleterious effect on cognitive development.[12] This finding suggests that the impact of seizure frequency on cognitive development may be age-dependent. Results have also been mixed for duration of seizure disorder.[10,11]

A few studies have compared the results of imaging studies to cognition function, and these studies have consistently found a relationship. Cognitive impairments have been associated with bilateral brain involvement,[5] asymmetry in cortical volume,[13] and reduced white matter volume.[10] Results of studies evaluating the significance of hemiparesis for general cognitive development have been inconsistent.[3,12,14]

Emotional Disturbance and Disruptive Behavior

Individuals with Sturge-Weber syndrome are clearly at higher risk for affective and behavioral disturbances than are individuals in the general population. Most of the early evidence was anecdotal. Patients were described who had temper outbursts, were irritable, or had other types of behavior problems.[15] Psychiatric problems in three Chinese patients were reported, two of whom presented with paranoid disorders and the third with pseudodementia.[16] Sujansky and Conradi were the first to describe mood and behavioral functioning in a larger sample of adults with Sturge-Weber syndrome, and reports of depressed mood and aggressive behavior were fairly common in their sample.[2]

Chapieski and colleagues assessed behavior and mood in a sample of young people with Sturge-Weber syndrome using standardized parent and

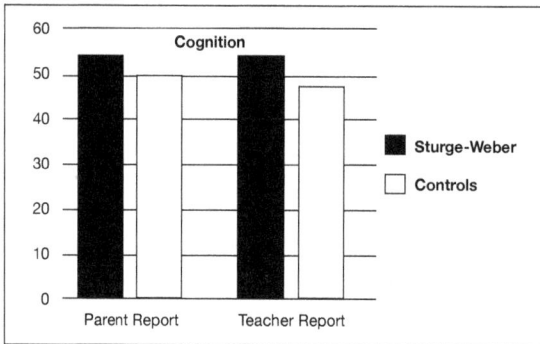

Table 5-1. Parent and teacher report of cognition and academic problems for a group of children with SWS without a seizure disorder and a group of demographically similar controls. Scales have a mean of 50 and a standard deviation of 10. Higher scores reflect more problems.

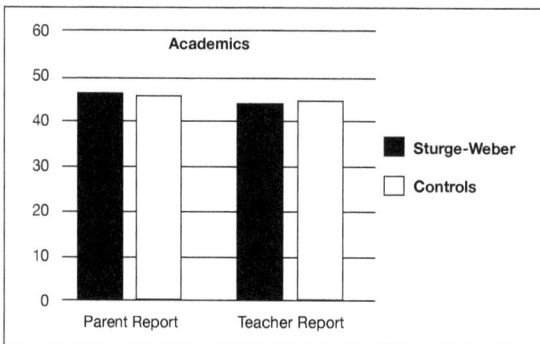

teacher questionnaires and a control group.[3] Consistent with the findings of the Sujansky and Conradi in adults, parents and teachers reported higher levels of mood disturbance in young people with Sturge-Weber syndrome. Parent reports indicated that young people with Sturge-Weber syndrome had more somatic complaints and exhibited more behaviors associated with depressed mood than their siblings. Twenty eight percent had depression scale scores in the clinically significant range and 42% scored in the abnormal range on a scale of somatic complaints. Parent reports of social withdrawal and anxiety, however, were not more common in the Sturge-Weber syndrome group than in the siblings. Teachers, similarly, reported more behaviors associated with general emotional distress in the Sturge-Weber syndrome group.

The participants in Sujanksy and Conradi's study reported an increased frequency of aggressive and noncompliant behaviors.[2] Teachers in the Chapieski et al. survey[3] similarly reported a higher incidence of oppositional behavior, conduct problems and behaviors associated with attention-deficit/hyperactivity disorder (ADHD) in the young people with Sturge-Weber syndrome than in their siblings, although the parents did not report a higher incidence of these disruptive behaviors. A more recent analysis of these data indicated that, within the Sturge-Weber syndrome group, 13% of the scores on the hyperactivity scale and 19% of the scores on the scale associated with noncompliance were elevated into clinically significant levels. Moreover, although parental responses to the behavior questionnaire suggested a similar frequency of hyperactive behavior among children with Sturge-Weber syndrome and their unaffected siblings, 22% of the children with Sturge-Weber syndrome had a diagnosis of ADHD, a much higher incidence than occurs in the general population.[17] Twenty-two percent of the teacher responses to the ADHD scale yielded scores in the clinically significant range, a number consistent with the percentage of the sample with Sturge-Weber syndrome who had an established diagnosis of ADHD. A recent report of 55 Sturge-Weber patients identified an even higher ADHD incidence of 42%.[9] Thus ADHD appears to be a common co-morbid condition of Sturge-Weber syndrome.

Risk Factors for Emotional Disturbance and Disruptive Behavior

Although Sujanksy and Conradi's survey participants reported a high incidence of depressed mood, the reports occurred more often in those with normal intelligence.[2] Chapieski and associates, conversely, found a negative correlation between depressed mood and level of intellectual functioning in their sample of children and adolescents.[3] The frequency of psychosomatic complaints was also inversely correlated with level of intellectual functioning. Consistent with the parent reports, teachers reported higher levels of emotional distress in children who had lower intellectual function. The discrepancy in findings between the two studies may reflect the age differences in the two samples or differences in assessment techniques. The Sujanksy-Conradi study was a retrospective report of adults and the occurrence of depressed mood was assessed with the use of a single question. The results may also be affected by the range of IQ levels in a particular sample since the assessment of mood in individuals with poor intellectual function can be difficult. Increased emotional difficulty in individuals with lower levels of intellectual functioning, however, would not be surprising given that this pattern has also been reported for individuals without Sturge-Weber syndrome.[18]

Chapieski and associates also suggest that the presence of epilepsy and the seizure frequency appear to increase the risk for emotional distress in individuals with Sturge-Weber syndrome, similar to findings in the general epilepsy literature.[3,19,20] These data also indicate that higher levels of family stress increased the likelihood of mood disturbance in children with Sturge-Weber syndrome, consistent with findings in individuals without Sturge-Weber syndrome. The presence of hemiparesis did not affect mood in the Chapieski et al. study, a somewhat surprising finding given that motor disorders in the pediatric population have been associated with social and emotional problems.[21]

Seizures may be a particularly important variable in identifying those individuals with Sturge-Weber syndrome who are at risk for emotional problems. Raches found no differences in parent-reported symptoms of emotional distress when a group of children with Sturge-Weber syndrome without a seizures were compared to a demographically similar normal control group.[8] Although the parents of children with Sturge-Weber syndrome and epilepsy reported more symptoms of emotional distress than those parents of affected children without epilepsy, the children with Sturge-Weber syndrome and epilepsy could not be differentiated from another group of intellectually similar children with epilepsy.

Sujanksy and Conradi noted that survey participants with lower levels of intellectual functioning were more likely to exhibit aggressive and oppositional behavior.[2] Similarly, teachers describe a higher frequency of oppositional and conduct problems in their students with Sturge-Weber syndrome who had lower intellectual function.[3] Lower levels of intellectual functioning, however, were not associated with a higher frequency of ADHD behaviors. Neither the presence of a hemiparesis nor a relatively high level of family stress correlated with teacher reports of any type of disruptive behavior.

Social Problems

Social problems may stem from inadequate social skills, negative reactions from the social environment, or some combination of the two. Sturge-Weber syndrome is associated with a number of medical conditions that could either inhibit the development of social skills or elicit negative reactions from others. Seizures, motor problems, impaired intellect, and port-wine stains have all been shown to alter social functioning in individuals without Sturge-Weber syndrome.[18,24,29-30] In our survey, both parents and teachers felt that the young people with Sturge-Weber syndrome were more likely to have impaired social skills than their siblings.[3] Forty-three percent of parents with children with Sturge-Weber syndrome reported significant social problems. Both parents and teachers believed that the presence of epilepsy increased the incidence of social dysfunction. Family stress increased the likelihood that parents would report social problems, and teachers reported more social problems in their students with Sturge-Weber syndrome who had hemiparesis. Level of intellectual functioning was inversely correlated with the number of social problems reported by parents but, surprisingly, not with social problems reported by teachers. This discrepancy might reflect the fact that the reference group for parents and teachers is different. Parents presumably compare their child with Sturge-Weber syndrome to their siblings or other normal children. In contrast, the comparison group for teachers of lower functioning students with Sturge-Weber syndrome might consist largely of other special education students. Social problems were highly correlated with mood problems by both parents and teachers. Because the data are correlational, it is unclear whether the social problems were the cause or the result of the emotional distress.

The term *adaptive skills* refers to an individual's level of independence with daily self-care tasks, functional communication, and social interactions and relationships. Measures of adaptive functioning provide a global assessment of whether an individual is functioning within the home and the larger community at an age-expected level and are, therefore, important measures of outcome. Unfortunately, the only data concerning adaptive functioning in individuals with Sturge-Weber syndrome came from a single study.[14] The level of adaptive functioning of children and adolescents in this study was below average, as was the average level of intellectual functioning. Adaptive functioning, however, was unrelated to level of intelligence. The single variable that predicted adaptive functioning was the severity of hemiparesis. Adaptive skills were unaffected by seizure frequency or the presence of a visual field deficit.

Although there are only limited data on the social and emotional consequences of port-wine stains in individuals with Sturge-Weber syndrome, there is a larger literature on the impact of port-wine stains in the general population. Patients and their families report that port-wine stains contribute to poor self-esteem, social discomfort in school, negative reactions from strangers, and reduced frequency of social interactions. [22,23,24] Social functioning sometimes improves after the port-wine stain has been treated with pulsed dye laser.[23] The social problems reported by parents and teachers were unrelated to the size of the port-wine stain but, as will be discussed below, the impact of the port-wine stain on social and emotional functioning appears to be age-related.[3]

Impact of Age on Cognitive and Behavioral Functioning

The lack of longitudinal data prohibits definitive conclusions about the natural history of cognitive and behavioral changes in individuals with Sturge-Weber syndrome. Cross-sectional data, nevertheless, can provide useful information about age-related changes. Chapieski and colleagues concluded that the age of the child was not significantly related to either parent or teacher reports of intellectual, developmental, or academic dysfunction.[3] Neither was there a significant association between age and assessed level of intellectual functioning. This lack of correlation between age and cognitive function suggests that general cognitive development in children with Sturge-Weber syndrome, at least during the school-age years, proceeds at a rate similar to that of children in the general population.

The likelihood of emotional or social deterioration, however, appears to increase with age. Both parents and teachers are more likely to report social problems in older children with Sturge-Weber syndrome.[3] Similarly, mood disorders seem to be more common in older children. A further analysis indicated that larger port-wine stains correlate with parent reports of depression and teacher reports of emotional distress and social dysfunction, but only for children over the age of 9 years. Together these findings suggest that younger children with Sturge-Weber syndrome may not be as sensitive as older children about their appearance or may be less likely to be teased by their young peers. This finding is consistent with studies of normal development that indicate that older children are more likely to compare themselves to others when evaluating themselves.[25] The increased sensitivity of older children and adolescents could contribute to depressed mood and increased social problems.

Conclusions and Future Directions

Although existing data are limited, the evidence of increased risk of cognitive, academic, mood, behavior, and social problems in individuals with Sturge-Weber syndrome is clear. The exact factors that place particular individuals with Sturge-Weber syndrome at risk for specific types of functional problems, however, are not always clear. In general, it appears that individuals with Sturge-Weber syndrome who have lower levels of intellectual functioning and those with epilepsy are more likely to have a broad range of emotional and behavioral problems. The small number of individuals with Sturge-Weber syndrome who do not develop seizures have a similar risk of intellectual impairment and emotional problems as individuals without Sturge-Weber syndrome. There is considerable variability of functional status, however, even in individuals with Sturge-Weber syndrome who do develop seizures.

Because epilepsy by itself is associated with a higher incidence of cognitive and behavioral problems, it is not surprising that seizures increase the risk of dysfunction in individuals with Sturge-Weber syndrome. A number of variables increase the risk for mood and behavioral abnormalities in individuals with epilepsy. Among patients with epilepsy but not Sturge-Weber syndrome, early age of seizure onset is a poor prognostic indicator for development.[19,26] Larger cohorts with Sturge-Weber syndrome support this relationship, although seizures typically begin in the first year of life in these individuals.[1,3] The few studies that explore the effect of seizure frequency, duration, and other variables yield inconsistent results, no doubt reflecting the small sample size of most Sturge-Weber syndrome cohorts.

The small number of imaging studies with Sturge-Weber syndrome have consistently demonstrated that the extent of brain involvement is correlated with the level of mental impairment. The effect of anticonvulsant agents on cognition and behavior has been well documented in the general epilepsy literature,[27,28] but the degree to which medication effects contribute to the dysfunction in individuals with Sturge-Weber syndrome is unknown. Finally, seizures and other chronic medical conditions can alter or restrict interactions within social environments, but little is known about the social interaction of individuals with Sturge-Weber syndrome within their families or in the larger community.[30]

Not only are there gaps in our understanding of the specific factors that create cognitive and behavioral dysfunction in individuals with Sturge-Weber syndrome, but we have only general descriptions of the problems themselves. Cognitive impairment, in particular, has not been well characterized. Studies have emphasized performance on measures of intelligence, so little is known about the range of more specific cognitive or academic impairments that occur in individuals with Sturge-Weber syndrome without global intellectual impairment. The reliance on global measures of cognition is understandable given that most studies have collected data by survey or chart review. Future prospective studies with detailed neuropsychological assessment are needed to clearly describe specific cognitive impairments associated with this disorder. Correlation of specific cognitive impairments with the results of brain imaging studies would, in addition, help identify the individuals with Sturge-Weber syndrome who are most likely to experience specific types of cognitive and learning problems.

With the exception of ADHD, which occurs more often in individuals with Sturge-Weber syndrome than in the general population, there are few data concerning the incidence of actual psychiatric disorders. The collection of behavioral data with questionnaires has confirmed that disagreeable behavior and negative moods are more common in individuals with Sturge-Weber syndrome, but we do not know how many would meet formal criteria for a psychiatric diagnosis or what the range of those diagnoses might be. We know even less about the extent to which these problems persist into the adult years because there are neither studies of psychiatric disorders or information about mood and behavior that has been collected with standardized instruments during the adult years.

There are currently no published longitudinal studies of Sturge-Weber syndrome, so little is known about the natural history of its effect on cognition and behavior. Available cross-sectional data suggest that cognitive development during the school-age years may be relatively stable, although the information is limited by the reliance on global assessments of intelligence and parent and teacher reports. More detailed assessments, however, might reveal subtle but consistent cognitive changes that are not apparent with more global measures.

Although there is little evidence that most school-aged children or adolescents regress cognitively, it appears that the likelihood of emotional and social dysfunction increases during middle childhood. The increased susceptibility to mood and social problems probably reflects the increasing sensitivity of most children to individual differences as they move beyond the early grades. One of the sources of sensitivity for young people with Sturge-Weber syndrome appears to be the port wine stain. Older children with larger port wine stains are more likely to experience depressed mood and social problems. There is some indication from the general literature on port wine stains that socioemotional functioning improves after laser treatment but the social impact of laser treatment has not been studied in patients with Sturge-Weber syndrome. Hemiparesis may increase the likelihood of social problems and impair adaptive functioning. Because limitations in independence would have greater implications for older children and adolescents than for younger children, the impact of a hemiparesis on emotional functioning may also be age-dependent.

Specific vulnerabilities of the elderly with Sturge-Weber syndrome have not been investigated. It is not known whether age-related deterioration in memory and other cognitive functions is more striking in individuals with Sturge-Weber syndrome. A more pronounced functional decline in older individuals with Sturge-Weber syndrome might reflect an exacerbation of the underlying neuropathological process that is specific to Sturge-Weber syndrome or an inadequate cognitive reserve resulting from brain insults earlier in life.[31] Longitudinal studies might help to determine whether cognition changes in near the end of the life span deviate from the expected.

Because severe intellectual impairments are common with this Sturge-Weber syndrome, it is easy to lose sight of the fact that many individuals with this condition live independent and highly productive lives despite the presence of epilepsy or other manifestations. The variability in psychosocial functioning is due, in part, to variability in the severity of the underlying neuropathology, but individual differences may also reflect differences in mediating factors such as social supports, therapeutic interventions, and educational programs. Identification of the most beneficial mediating factors will be important if quality of life in individuals with Sturge-Weber syndrome is to be enhanced.

REFERENCES

1. Sujansky E, Conradi S: Sturge-Weber Syndrome: Age of onset of seizures and glaucoma and the prognosis for affected children. *J Child Neurol* 1995;10:49-58.

2. Sujansky E, Conradi S: Outcome of Sturge-Weber syndrome in 52 adults. *Am J M Genet* 1995;57:35-45.

3. Chapieski L, Friedman A, Lachar D: Psychological functioning in children and adolescent with Sturge-Weber syndrome. *J of Child Neurol* 2000;15:660-665.

4. Pascual-Castroviejo I, Diaz-Gonzales C, Garcia-Melian R, et al: Sturge-Weber syndrome: Study of 40 patients. *Pediatr Neurol* 1993;9:283-288.

5. Bebin E, Gomez M: Prognosis in Sturge-Weber disease: Comparison of unihemispheric and bihemispheric involvement. *J Child Neurol* 1988;3:181-184.

6. Bennett T, Maile R. The neuropsychology of pediatric epilepsy and antiepileptic drugs. In: Reynolds C, Fletcher-Jantzen E, eds. *Handbook of Clinical Child Neuropsychology*. New York: Plenum Press, 1977;517-538.

7. Fastenau P, Jianzhao S, Dunn D, et al: Academic underachievement among children with epilepsy: proportion exceeding psychometric criteria for learning disability and associated risk factors. *J Learn Disabil* 2008;41:195-207.

8. Raches D: Differentiating the effects of epilepsy in children with Sturge-Weber Syndrome. Unpublished master's thesis, University of Houston, Houston, Texas, 2006.

9. Pascual-Castroviejo I, Pascual-Pascual SI, Velazquez-Fragua R, et al: Sturge-Weber syndrome: study of 55 patients. *Can J Neurol Sci* 2008;35:301-307.

10. Juhász C, Lai C, Behen M, et al: White matter volume as a major predictor of cognitive function in Sturge-Weber Syndrome. *Arch Neurol* 2007;64:1169-1174.

11. Chapieski L, Raches D, Hiscock M et al: Cognitive impairments in children with Sturge-Weber syndrome: is it just the seizures? *Epilepsia* 2006;47:108.

12. Kramer U, Kahana E, Shorer Z, et al: Outcome of infants with unilateral Sturge-Weber Syndrome and early onset seizures. *Develop Med Child Neurol* 2000;42:756-759.

13. Kelley TM, Hatfield LA, Lin DD, et al. Quantitative analysis of cerebral cortical atrophy and correlation with clinical severity n unilateral Sturge-Weber syndrome. *J Child Neurology* 2005;20:867-870.

14. Reesman, J, Gray R, Suskauer SJ, et al: Hemiparesis is a clinical correlate of general adaptive dysfunction in children and adolescents with Sturge-Weber Syndrome. *J Child Neurology* 2009;34:701-708.

15. Falconer M, Rushworth R: Treatment of encephalotrigeminal angiomatosis (Sturge-Weber Disease) by hemispherectomy. *Arch Dis Child* 1960;35:443-447.

16. Lee S: Psychopathology in Sturge-Weber syndrome. *Can J Psychiatry* 1990;35:674-678.

17. American Psychiatric Association: *Diagnostic and Statistical Manual of Mental Disorders, 4th ed.* Washington, DC, American Psychiatric Association, 1994.

18. Szymanski L: The retarded child and adolescent, in Kestenbaum C, Williams, D (eds): *Handbook of Clinical Assessment of Children and Adolescents*, vol I. New York, Springer-Verglag, 1992, 109-136.

19. Hamiwka LD, Wirrell EC: Comorbidities in pediatric epilepsy: beyond "just" treating the seizures. *J Child Neurol* 2009;24:734-742.

20. Turky A, Beavis JM, Thapar AK, et al: Psychopathology in children

and adolescents with epilepsy: an investigation of predictive variables. *Epilepsy Behav* 2008;12:136-144.

21. Dewey D, Cawford SG, Wilson, BN, et al: Co-occurrence of motor disorders with other childhood disorders in Dewey D, Tupper, D (eds): *Developmental Motor Disorders: a Neuropsychological Perspective*, The Guilford Press, New York, 2004, 405-426.

22. Lanigan SW, Cotterhill, JA: Psychological disabilities amongst patients with port wine stains. *Br J Dermatol*, 1989;121:209-215.

23. Troilius A, Wrangsjo B, Ljunggren B: Potential psychological benefits from early treatment of port-wind stains in children. *Br J Dermatol*, 1998;139:59-65.

24. van der Horst CM, DeBorgie CA, Knopper JL, et al: Psychosocial adjustment of children and adults with port wine stains. *Br J Plast Surg*, 1997;50:463-467.

25. Harter S: The developmental of self-representation, in Damon W, Eisenberg N (eds): *Handbook of Child Psychology, 5th ed, vol 3, Social, Emotional and Personality Development*. New York, John Wiley & Sons, 1998, 553-618.

26. Motamedi G, Meador K: Epilepsy and Cognition. *Epilepsy Behav*, 2003;4 Suppl 2:s25-38.

27. Loring D, Meador K: Cognitive and behavioral effects of epilepsy treatment. *Epilepsia*, 2001;42:24-32.

28. Goldberg J, Burdick K. Cognitive side effects of anticonvulsants. *J Clin Psychiatry*, 2001;62:27-33.

29. Larkin D, Summers J: Implications of movement difficulties for social interaction, physical activity, play, and sports in Dewey D, Tupper, D (eds): *Developmental Motor Disorders: a Neuropsychological Perspective*, The Guilford Press, New York, 2004, 443-460.

30. Chapieski L, Brewer V, Evankovich K, et al: Adaptive functioning in children with epilepsy: contributions of maternal anxiety

31. Satz P: Brain reserve capacity on symptom onset after brain injury: a formulation and review of evidence for threshold theory. *Neuropsych*, 1993;7:273-295.

Chapter 6

Imaging Brain Structure and Function in Sturge-Weber Syndrome

Csaba Juhász, M.D., Ph.D.

Bálint Alkonyi, M.D.

Harry T. Chugani, M.D.

Neuroimaging studies, particularly magnetic resonance imaging (MRI) with sequences optimized for the detection of abnormalities associated with Sturge-Weber syndrome, are instrumental in the initial diagnosis, evaluation of the extent and severity of intracranial involvement, and also in detection of progressive cerebral changes during the course of the disease. Both structural and functional neuroimaging modalities play an important role in the presurgical evaluation of patients with Sturge-Weber syndrome and intractable seizures. Multimodality imaging studies with combination of MRI and functional imaging (such as positron emission tomography [PET]) can reveal structural and functional correlates of neuro-cognitive abnormalities in Sturge-Weber syndrome. Imaging studies are also invaluable in providing insights into the pathophysiology and mechanism of progression of Sturge-Weber syndrome.

In this chapter, we provide an overview of both conventional and advanced imaging modalities used in the radiological investigation of Sturge-Weber syndrome. It should be noted that conventional distinctions between "structural" and "functional" neuroimaging techniques are diminishing, as traditional structural neuroimaging techniques, such as various MRI methods, are increasingly capable of characterizing functional aspects (e.g., blood flow or biochemical changes); similarly, traditionally "functional" techniques, such as PET, have now excellent anatomical resolution, thus providing details of structural abnormalities in addition to brain function. The combination of these approaches provides previously unattainable insights into details of tissue pathology and functional changes. Clinical uses and limitations of each imaging modality, in the context of imaging Sturge-Weber syndrome, are discussed below.

X-ray and Cerebral Angiography

Historically, the radiologic method used to establish the diagnosis of Sturge-Weber syndrome was the skull X-ray, which classically showed "tram track-like" calcifications.[1] These, however, are often not present in young children with Sturge-Weber syndrome, and plain X-ray is not capable of directly visualizing brain tissue or vessel abnormalities; therefore, X-ray is not used any more in the diagnostic work-up of Sturge-Weber syndrome. Cerebral angiography in patients with Sturge-Weber syndrome can reveal the variable presence of arterial thromboses, absence or paucity of cortical veins, aberrant cerebral venous sinus drainage and arteriovenous malformations, in addition to the characteristic leptomeningeal angioma, most commonly located in the posterior parietal distribution.[2, 3] However, angiography is often not sufficient to demonstrate the true extent of the vascular malformation, which itself may be difficult to demonstrate due to low flow and minimal accumulation of the angiographic contrast agent. Nevertheless, angiographic findings, together with those from conventional brain scanning with (99m)Tc pertechnetate,[4] have suggested that unilateral chronic cerebral ischemia may be at least partially responsible for the pathophysiological processes of cerebral calcification and cell death in Sturge-Weber syndrome. Conventional catheter angiography is an invasive procedure and nowadays it is not considered to be a part of the radiological work-up in patients with suspected Sturge-Weber syndrome.

Computed Tomography (CT)

Calcification, which becomes more obvious with age in patients with Sturge-Weber syndrome, can be demonstrated reliably with cranial computed tomography (CT).[5] The calcifications typically have a gyriform shape on the scans (**Figure 6-1**). Sometimes microcalcifications can also show up as a diffuse increased density on the images. Calcified areas may not be present at birth or in young infants; rather, they can appear during the course of the disease and progress slowly on serial images. Gyriform calcification is not specific for Sturge-Weber syndrome: similar findings can be seen in some cases with chronic cerebral infarction, meningitis, or celiac disease. Indeed, celiac disease

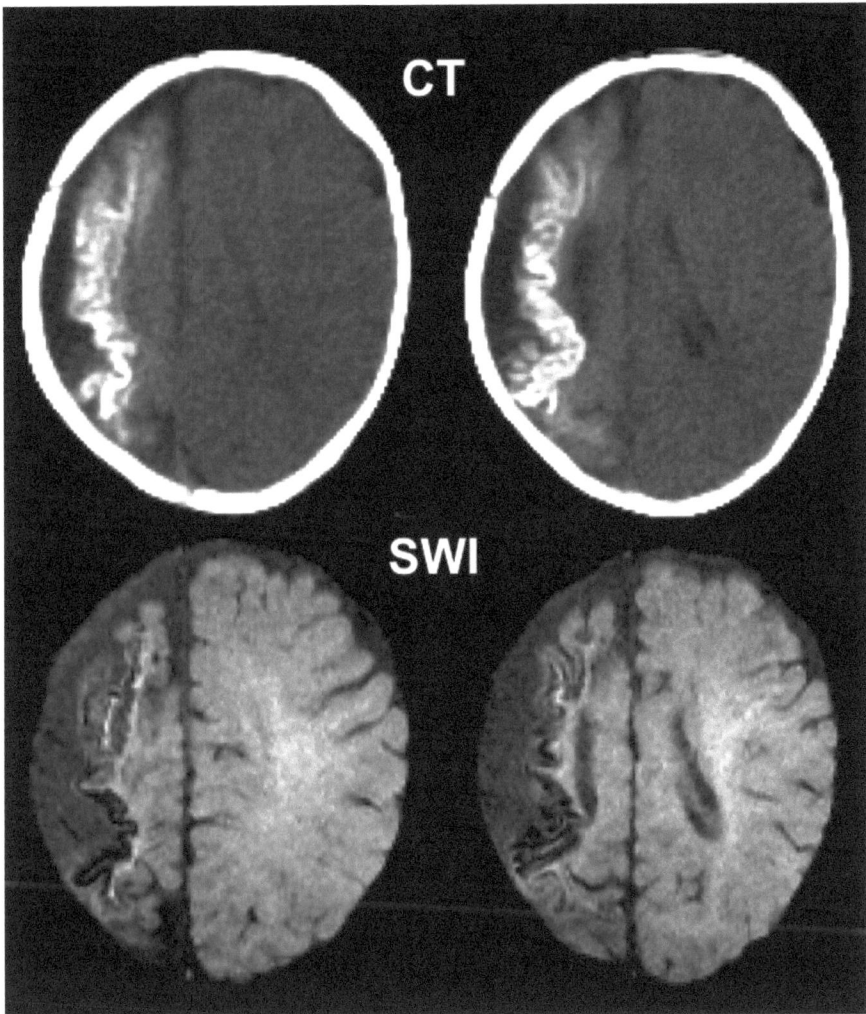

Figure 6-1. Axial CT images and corresponding axial planes of MR susceptibility weighted imaging (SWI) in a 3-month-old infant with Sturge-Weber syndrome and extensive right hemispheric involvement. The CT scan shows right hemispheric atrophy with gyriform calcification in the frontoparietal regions. SWI images also visualize calcified regions, which are seen as low signal intensity (black) areas following the cortical mantle. Also note that the configuration of the cortex is consistent with polymicrogyria, which was verified on histopathology examination after resective surgery. The child underwent hemispherectomy at 6 months of age because of medically refractory seizures. He became seizure-free and is developing well more than 3 years after surgery, with minimal weakness in the left extremities.

should be always in the differential diagnosis in children with parieto-occipital calcifications not associated with a facial port wine stain.[6] These calcifications are not accompanied by typical leptomeningeal enhancement seen in Sturge-Weber syndrome.

Contrast-enhanced CT scan may show ipsilateral choroid plexus enlargement, abnormal draining veins, and signs of blood-brain barrier breakdown due to seizures. CT with contrast injection can be done in the emergency setting after a first seizure in a child with a facial port wine stain. Extensive brain atrophy is also often apparent with CT, mostly in advanced cases, but subtle atrophy can be more readily demonstrated by MRI.[7,8] CT can also be used to demonstrate overgrowth of the calvaria (mostly in the form of frontal bone hypertrophy associated with frontal sinus enlargement) and skull base that can accompany brain atrophy. A case report demonstrated that brain CT angiography can detect a small angiomatous nidus (which led to a rare hemorrhage in that case) despite previously normal post-contrast MRI and CT scans.[9] Overall, CT imaging can demonstrate several features of intracranial involvement, but a normal CT scan does *not* exclude Sturge-Weber syndrome, as several abnormalities associated with this disorder are more reliably depicted by MRI than by CT imaging.

Magnetic Resonance Imaging (MRI) Techniques

MRI is currently the single most important imaging modality for Sturge-Weber syndrome. For clinical evaluation, conventional MR imaging protocols should include native, spin-echo T1-weighted and T2-weighted images, as well as post-gadolinium T1-weighted images as a minimum. Gradient-recalled-echo (GRE) images are also advised to depict calcifications. It must be noted that MRI techniques have been evolving rapidly, and development of newer, advanced MRI sequences constantly improves our ability to detect vessel and tissue abnormalities as well as functional changes associated with Sturge-Weber syndrome. Thus, optimal MR imaging of patients with Sturge-Weber syndrome increasingly incorporates some of the advanced MRI techniques, including some that allow quantification of various aspects of cerebral abnormalities (e.g., those associated with altered water diffusion, perfusion, or blood oxygenation), thus providing more objective measures of brain damage. The clinical uses and limitations of the most important conventional and advanced MRI sequences and modalities used in Sturge-Weber syndrome are detailed below.

T1-Weighted Post-Gadolinium MRI. The conventional "gold standard" for radiological diagnosis of Sturge-Weber syndrome is a T1-weighted MRI with gadolinium contrast, which delineates the leptomeningeal angioma, the hallmark of intracranial involvement (**Figure 6-2**).[10-13] This abnormality is most commonly seen over the posterior parietal, occipital, and temporal lobes, less frequently affecting the frontal lobe; the leptomeningeal angioma can also affect the entire hemisphere and, in some cases, can be confined only to the frontal lobe. In rare cases, angiomatous changes can extend to the infratentorial compartment, and cases with exclusively infratentorial involvement have also been described.[14] Lack of leptomeningeal angiomatosis in the presence of deep venous occlusion and frontal collaterals can also occur.[15]

Post-gadolinium MRI can also demonstrate other common vessel abnormalities associated with Sturge-Weber syndrome, including enlarged choroid plexus and transmedullary as well as subependymal veins (**Figure 6-2**). The superficial and deep venous systems are connected by anastomotic veins running through the centrum semiovale. In a normal brain a venous watershed exists between the deep and superficial venous systems, but the significance of these anastomoses is marginal.[16,17] In Sturge-Weber syndrome,

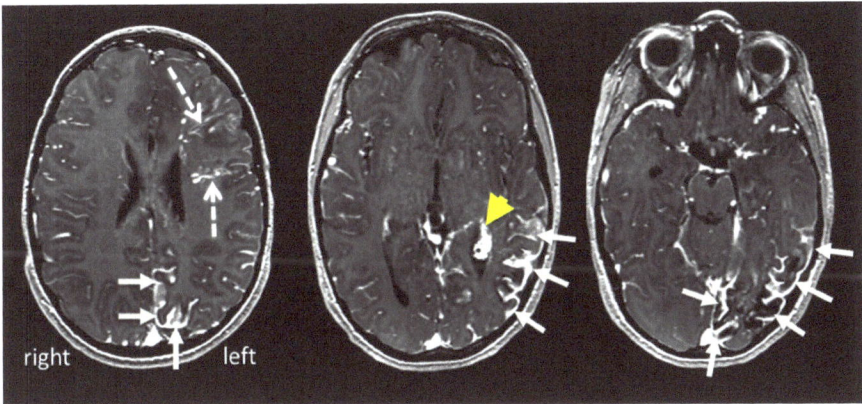

Figure 6-2. Axial, T1-weighted, gadolinium-enhanced MR images of a 6-year-old girl with Sturge-Weber syndrome and left hemispheric involvement. The images demonstrate three typical intracranial vessel abnormalities: 1. leptomeningeal angioma in the posterior parietal, temporal, and occipital regions (solid arrows)—this is the hallmark of intracranial involvement in Sturge-Weber syndrome; 2. enlarged choroid plexus (arrowhead); 3. deep transmedullary veins in the frontal lobe white matter (dotted arrows)—these veins drain blood into subependymal veins located in the wall of the frontal horn of the ventricles. In addition, moderate atrophy is seen in the areas underlying the vessel malformation.

these deep medullary veins are often enlarged and drain venous blood from the cortex with insufficient superficial venous drainage centripetally into the deep (galenic) venous system. Thus, deep veins in Sturge-Weber syndrome are likely collateral pathways supporting venous drainage to compensate for the lack of normal venous drainage through cortical veins. Interestingly, deep transmedullary veins, passing through the white matter, can sometimes be seen in the frontal lobe of patients with posterior leptomeningeal involvement. In many cases, however, this compensatory deep venous drainage eventually becomes insufficient, leading to decreasing cortical metabolic activity due to venous stasis, decreased arterial pressure and resulting hypoxia.

Enlargement of the choroid plexus, typically ipsilateral to the leptomeningeal angioma, can be well demonstrated by post-gadolinium MRI (**Figure 6-2**). The size of the choroid plexus is positively correlated with the extent of leptomeningeal involvement.[18] It is likely that enlargement of the choroid plexus, with its thin-lining walls and sponge-like characteristics, reflects increased pressure in the deep venous system.[3, 18]

Despite its superior ability to depict the above-detailed vessel abnormalities in Sturge-Weber syndrome, T1-weighted, post-gadolinium MRI has limitations. First, gadolinium-enhanced MRI may be occasionally negative during the first few months (or even years) of life despite subsequent demonstration of radiological signs of Sturge-Weber syndrome.[14,19] To minimize this pitfall, one can include additional MRI sequences such as susceptibility-weighted imaging (SWI) or perfusion weighted imaging (PWI) in the MRI protocol of infants with suspected Sturge-Weber syndrome (see below for more details on these techniques). Post-contrast fluid-attenuated inversion recovery (FLAIR) imaging may also have superior sensitivity for detecting the leptomeningeal angioma, probably because of suppression of signal intensity from normal vascular structures on the surface of the brain by FLAIR, thus allowing easier visualization of abnormal leptomeninges.[20] In cases of completely negative results, MR imaging can be repeated in a year or so if clinical suspicion of Sturge-Weber syndrome persists. Importantly, the family should not be given the false hope of absent leptomeningeal angioma based on an MRI performed in the first few months of life.

T1-weighted post-gadolinium MRI may sometimes miss a small leptomeningeal angioma even in older patients. Leptomeningeal enhancement may also not be present in the later stages of Sturge-Weber syndrome due to obliteration of the abnormal vessels. In these cases, other signs of venous involvement (such as deep venous abnormalities) can assist in the correct diagnosis.

Finally, transient leptomeningeal enhancement related to recent seizures also has been reported.[21] A similar phenomenon has been described on contrast enhanced CT scans.[22] These latter findings suggest that subacute hemodynamic changes and/or transient blood-brain barrier impairment associated with seizure activity can, at least in some cases, affect the radiological appearance of the angioma.

Gradient-Recalled Echo (GRE) Imaging. GRE images with a long echo time are quite reliable in detecting calcifications (**Figure 6-3**). Indeed, this technique is even superior to CT in the demonstration of microcalcifications.[7,23] Visualization of calcium-rich regions as marked hypointensities on GRE images can be due to T2* shortening from static local magnetic field gradients at interfaces of brain regions differing in magnetic susceptibility. This effect is further utilized in susceptibility weighted imaging (SWI), which is becoming an integral part of clinical MRI imaging of Sturge-Weber syndrome (see below).

T2-Weighted and FLAIR Imaging. T2-weighted MRI can reliably detect brain atrophy with enlargement of the subarachnoid space and ventricles (**Figures 6-3** and **6-4**). Frontal hyperostosis is also well visible on both T2 and

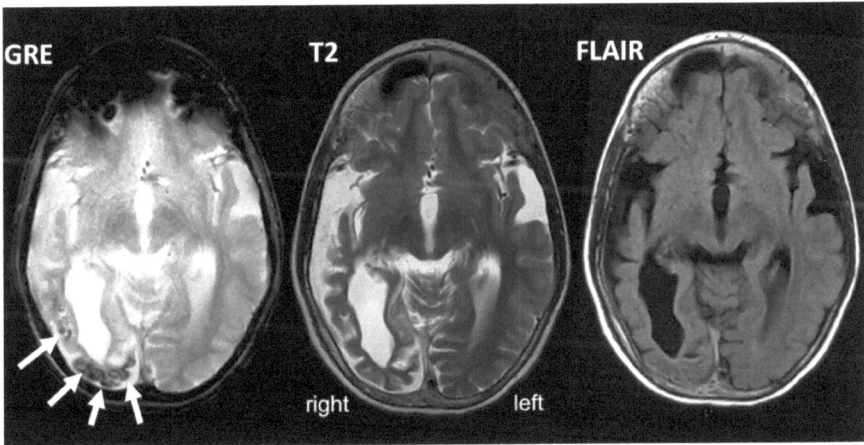

Figure 6-3. Axial gradient-recalled echo (GRE), T2-weighted (T2), and fluid attenuation inversion recovery (FLAIR) MR images of a 10-year-old boy with Sturge-Weber syndrome affecting the right temporo-occipital region. GRE images detected occipital gyriform calcification as low signal-intensity (dark) areas (arrows) in the posterior cortex. Corresponding T2 and FLAIR images demonstrate severe temporo-occipital atrophy but do not visualize calcifications.

T1-weighted images. A particularly useful aspect of T2-weighted MR imaging in Sturge-Weber syndrome is detection of "accelerated myelination" in the affected hemisphere, which is commonly seen in infants who present with early neurological symptoms.[24,25] Accelerated myelination can be manifested as a hypo-signal on T2 images, often associated with an increased T1 signal (signal inversion). Similar MRI findings have been described in a few children with cerebral venous sinus thrombosis not associated with Sturge-Weber syndrome.[26] This suggests that accelerated myelination may develop as a result of venous congestion and related hypoxia, not necessarily specific for Sturge-Weber syndrome. In addition, recent studies indicate that Sturge-Weber syndrome may be associated with previously overlooked cortical developmental malformations, manifested as increased cortical thickness, gray-white matter junction blurring, microgyria and T2/FLAIR (fluid attenuated inversion recovery) gray and white matter signal intensity changes.[27] These abnormalities are associated with histopathology findings of focal cortical dysplasia and polymicrogyria (see an example on **Figure 6-1**, where abnormalities consistent

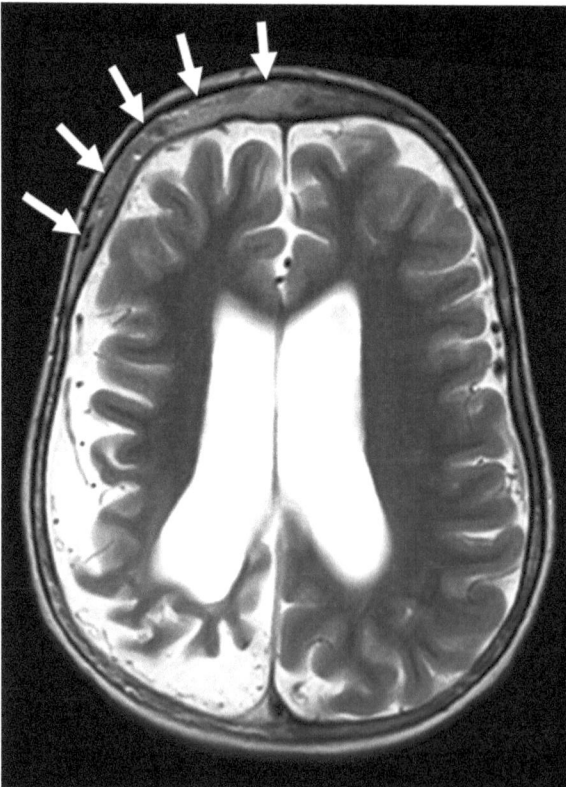

Figure 6-4. Axial T2-weighted image of a child with Sturge-Weber syndrome affecting the right hemisphere. The image demonstrates severe hemispheric atrophy on the right side with enlarged subarachnoid space and lateral ventricle. In addition, right hyperostosis is seen (arrows); this is most common in the frontal bone, even if atrophy is more severe in the posterior brain regions.

with polymicrogyria are seen on MR SWI images). Since these are potentially epileptogenic malformations, the findings suggest that such developmental malformations may contribute to focal seizures in Sturge-Weber syndrome. Therefore, MR imaging with multiple sequences should be utilized to carefully look for signs of cortical developmental malformations, which need to be considered during presurgical evaluation of patients with Sturge-Weber syndrome and intractable seizures.

Susceptibility Weighted Imaging (SWI). SWI utilizes the susceptibility differences between tissues to create a type of contrast using phase information to enhance the "native" contrast of tissues in magnitude images. Deoxyhemoglobin (paramagnetic) and calcifications (diamagnetic) cause small distortions of the local magnetic field and result in susceptibility changes. Abnormal venous oxygenation, calcification, micro-hemorrhages, or any combination could generate magnetic field susceptibility contrast in SWI. SWI is exquisitely sensitive to the venous vasculature by detecting deoxygenated blood in small veins without administration of a contrast medium.[28-30] Thus, SWI is an ideal MRI technique to detect fine deep venous abnormalities in children with Sturge-Weber syndrome non-invasively (without any contrast administration), even if conventional MRI shows no definite structural changes.[19,31] SWI has superior sensitivity to conventional T1-weighted gadolinium-enhanced MRI sequence by showing fine details of deep transmedullary and periventricular veins; maximum intensity projection (MIP) images of SWI acquisitions can demonstrate exquisite details of the abnormal deep venous network (**Figure 6-5**). SWI is also very sensitive to calcified gyriform abnormalities in Sturge-Weber syndrome (see **Figure 6-1**), while T1-weighted images are more sensitive to visualizing the choroid plexus and abnormal pial vessels. Thus, SWI and T1-weighted, gadolinium-enhanced MRI provide complementary information regarding brain abnormalities in Sturge-Weber syndrome[32] and, where available, SWI should be incorporated in clinical MRI protocols.

Perfusion Weighted Imaging (PWI). PWI estimates cerebral blood flow (CBF) by tracking a bolus of exogenous, non-diffusible, high magnetic susceptibility contrast agent (such as gadolinium-diethylenetriaminepentaacetic acid) using MRI, followed by tracer kinetic analysis to generate hemodynamic parameter maps. Detected signal changes are robust (10% to15%) with a single dose of gadolinium similar to perfusion CT. Quantitative maps of CBF, cerebral blood volume, mean transit time, and time to peak enhancement can be generated.

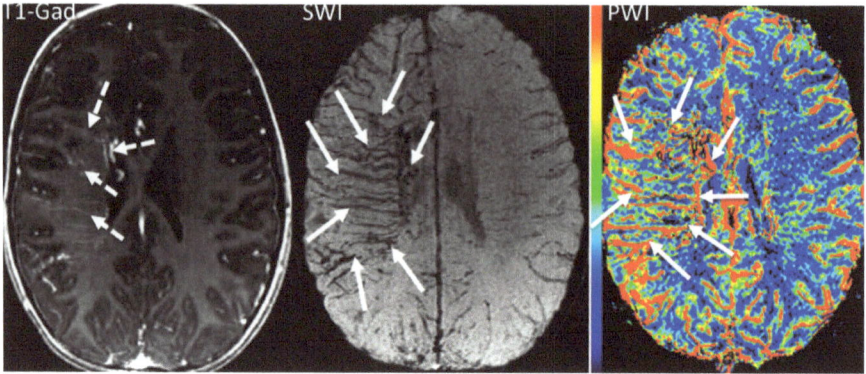

Figure 6-5. Comparison of axial T1-weighted post-gadolinium (T1-Gad), native (pre-gadolinium) susceptibility weighted imaging (SWI) and perfusion weighted imaging (PWI) of an 11-year-old girl with Sturge-Weber syndrome. All images demonstrate the presence of central transmedullary and subependymal veins in the right central region. SWI, despite lack of contrast administration, shows details of small vein abnormalities (arrows) better than the post-contrast T1-weighted image. PWI shows increased cerebral blood volume in the affected area.

PWI is ideally suited for children because of fast acquisition, repeatability, and its susceptibility for microvascular abnormalities. In stroke imaging, PWI is highly sensitive for detecting ischemic tissue at risk of infarction.[33] In Sturge-Weber syndrome, a few small series using PWI have demonstrated decreased perfusion in areas with meningeal enhancement and, in some cases, in adjacent areas (**Figure 6-6**).[34,35] Additional perfusion defects were seen in areas of deep venous abnormalities. As expected, gadolinium arrival in the affected areas was normal followed by a delayed clearance due to impaired venous drainage. Arterial perfusion deficits were only present in the most severely affected areas. The hypoperfused parenchyma volume showed a good correlation with motor deficits and some correlation with overall disability scores in Sturge-Weber syndrome.[35] Early perfusion changes were also demonstrated in a 9-month-old boy with Sturge-Weber syndrome and new onset seizures.[36] Thus, PWI can be a valuable clinical imaging approach to detect and quantify severity of early perfusion abnormalities related to neurological dysfunction. A more recent PWI study also reported *increased* regional perfusion in two children with recent breakthrough seizures;[37] this observation suggests that vasodilatation due to recent seizures or subacute ischemia can affect cerebral blood flow and perfusion, and this may need to be taken into account when interpreting PWI

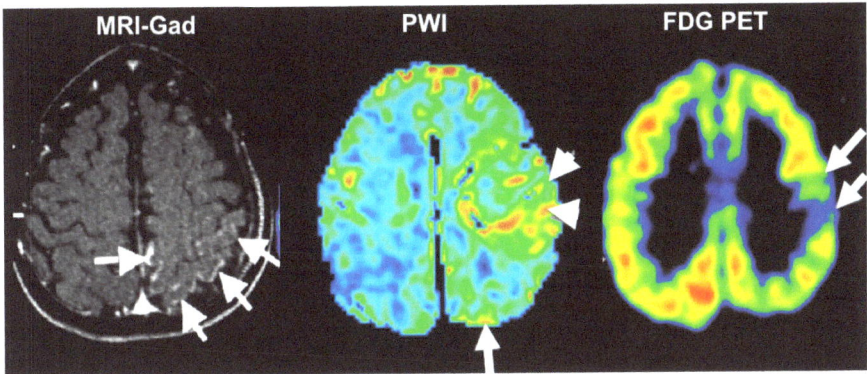

Figure 6-6. Comparison of abnormalities shown on gadolinium-enhanced MRI (MRI-Gad), perfusion-weighted imaging (PWI, a cerebral blood volume map is shown), and FDG PET. The figure demonstrates that perfusion abnormalities on the cerebral blood volume map, indicating increased blood volume, extend beyond the area of leptomeningeal enhancement (arrows), which is mild in this case. FDG PET shows severe hypometabolism (arrows) in the same central area where the perfusion abnormality was seen.

images in Sturge-Weber syndrome. Although spatial resolution of PWI images are currently limited, advances in image acquisition and processing techniques will likely continue to improve the quality and quantification techniques of perfusion imaging.

Diffusion-Weighted and Diffusion Tensor Imaging (DTI). Diffusion-weighted imaging (DWI) is an MRI application used for the evaluation of the microstructure of brain tissue. DWI generates image contrast based on the regional differences in diffusion of water molecules within the brain, allowing calculation of the apparent diffusion coefficient (ADC), which measures the magnitude of water diffusion in brain regions. With progressive myelination during brain development, the overall motion of water molecules decreases, and this leads to decreased ADC values. In contrast, ADC values may be altered (more often increased than decreased) in pathological processes affecting tissue integrity that occur as a result of various pathologic processes. Importantly, these pathological changes are not always detectable on conventional MRI.

A study of 15 patients with Sturge-Weber syndrome found increased ADC values in the normal-appearing white matter of the frontal, parietal, and occipital lobes and even in the pons.[38] In addition, ADC values of the cerebellar white matter were increased in six out of eight affected cerebellar lobes. These findings indicated that white matter diffusion abnormalities, suggestive of altered tissue integrity, may often extend beyond structural abnormalities seen on conventional MRI. Such widespread abnormalities may contribute to neurocognitive deficits. Indeed, initial diffusion tensor imaging (DTI) studies support this notion (see below).

DTI can assess not only the magnitude but also can quantify directionality of water diffusion (**Figure 6-7**).[39,40] Fractional anisotropy (FA) is an often-used diffusion parameter in white matter structures and is an indicator of white matter coherence, axonal organization, and myelination. DTI studies in Sturge-Weber syndrome have demonstrated diffusion changes in white matter extending beyond apparent cortical abnormalities and correlating with cognitive impairment.[31] Thus DTI may be a sensitive MRI method to detect and quantify early white matter involvement in Sturge-Weber syndrome, before macrostructural abnormalities become apparent.

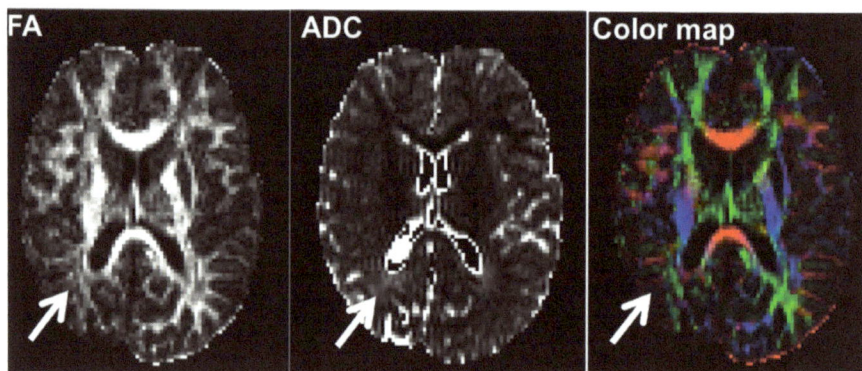

Figure 6-7. Diffusion tensor imaging (DTI) in Sturge-Weber syndrome (9-year-old boy with right posterior hemispheric involvement). Fractional anisotropy (FA) map showed decreased FA values in the parietal white matter (arrow). The same region also showed a mild increase of the apparent diffusion coefficient (ADC) values; both FA and ADC changes indicated impaired white matter integrity in this area. On the color vector map, fiber directions are color-coded based on the main direction of water diffusion (indicating the main fiber direction: red - left-right, green - antero-posterior, blue - superior-inferior). The map demonstrates loss of normal fiber directionality (which is mainly antero-posterior, i.e., green, on the intact side) caused by white matter damage in the affected posterior parietal area. The frontal regions appear symmetric on all three maps, indicating normal diffusion. The patient underwent right posterior resection and became seizure-free (>1 year follow-up).

Although further studies with larger patient samples are needed, DTI may also be useful to detect early white matter involvement, consistent with accelerated myelination, in newborns or infants with suspected Sturge-Weber syndrome during a period when other, conventional MRI techniques may not show definite abnormalities.[41] Cerebral diffusion changes can be also detected in deep grey matter structures such as the thalamus. A study of 20 children with Sturge-Weber syndrome and unihemispheric brain involvement showed increased mean diffusivity (a measure related to ADC) in the thalamus ipsilateral to the angioma,[42] while FA abnormalities were less robust. Interestingly, thalamic diffusivity and glucose metabolic abnormalities (measured by glucose PET) were highly correlated ($p<0.001$), indicating lower thalamic glucose metabolism associated with high diffusivity. In addition, thalamic asymmetries of diffusivity proved to be a good predictor of cognitive function (full scale IQ) in these children, and may prove to be a good imaging marker in this respect.

In addition to providing measurements of white matter diffusion in various regions, the DTI technique also allows us to study the integrity of specific white matter tracts using fiber tracking. This approach is increasingly used to delineate major fiber tracts, such as the corticospinal tract or speech related tracts (e.g., the arcuate fasciculus). In particular, DTI tractography provides useful information prior to resective brain surgery and can assist in preservation of vital tracts to avoid post-surgical functional deficits. It has been reported that the integrity of the corticospinal tract after stroke has important predictive value for recovery of motor functions in the post-stroke period.[43] A DTI study in children with unilateral Sturge-Weber syndrome evaluated the integrity of the corticospinal tract and showed diffusion abnormalities even before obvious motor symptoms occurred in some children.[44] Thus, DTI tractography of the corticospinal tract, and perhaps other major fiber tracts in the brain, may be sensitive measures to detect early changes in specific pathways prior to symptom onset.

MR Venography. MR venography is often used to diagnose cerebral venous thrombosis. The most commonly used MR venography techniques include unenhanced time-of-flight (TOF) MR venography as well as post-contrast MR venography (**Figure 6-8**). Two-dimensional TOF techniques are useful to evaluate the intracranial venous system because of their excellent sensitivity to slow flow and their diminished sensitivity to signal loss from saturation effects compared with the sensitivities of three-dimensional TOF techniques.[45] Venous flow in the plane of image acquisition may produce saturation and nulling of the venous signal, a potential pitfall for image

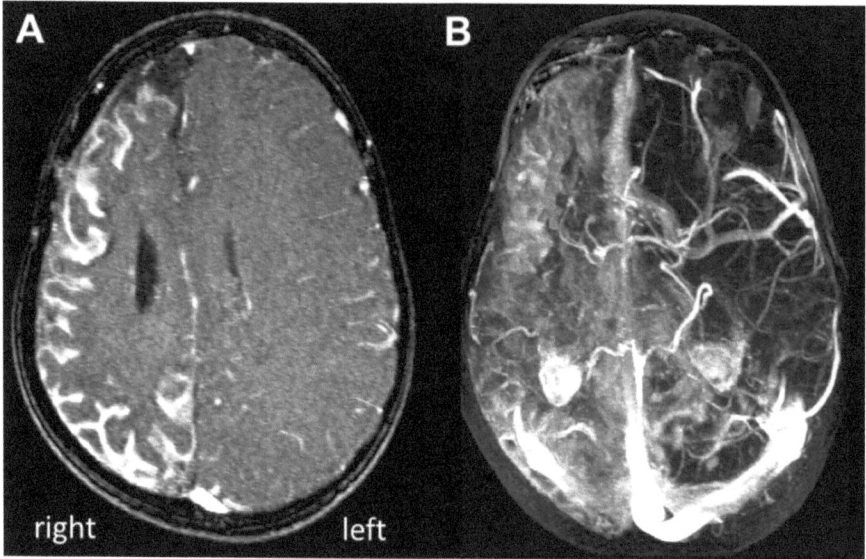

Figure 6-8. Post-contrast MR venography images of a 2 year- and 8-month-old boy with Sturge-Weber syndrome and extensive right hemispheric angioma. The pial vascular malformation is seen as a bright contrast enhancement on the time-of-flight source image (A). The entire venous system is visualized on the reconstructed maximum intensity projection (MIP) image (B), which shows normal veins and venous sinuses in the left hemisphere, while many of these are missing or poorly developed on the right side. Poor filling of the transverse sinus, prominent choroid plexus, and a hazy appearance of abnormal veins over most of the right hemisphere characterize the severe venous blood flow abnormalities on this side.

interpretation and diagnosis with TOF venography.[46] A close inspection of the source images is mandatory to accurately evaluate venous morphologic features and reduce the potential for diagnostic error. A few reports have demonstrated the potential usefulness of MR venography in Sturge-Weber syndrome. Vogl and colleagues[47] studied four children with Sturge-Weber syndrome and found reduced flow of the transverse sinuses and jugular veins along with a prominent deep collateral venous system and a paucity of superficial cortical veins. Arterial MR angiography was less instructive, although it revealed a reduced flow signal from the left middle cerebral artery in one hemiparetic patient and angiomatous changes of high branches of the middle cerebral artery in two children. Cure and colleagues[48] reported signs of progressive venous occlusion in an infant, whose initial MRI was performed three days after birth, followed by a second scan eight months later, when he started to have seizures. MR venography in this child demonstrated interval occlusion of the straight sinus and concomitant increased prominence of the falcine sinus, providing rare evidence for progressive venous sinus occlusion and redistribution of the venous flow in the early stage of the disease. Subsequently, Juhász and colleagues[49] reported the use of post-contrast TOF MR venography for superior visualization of a leptomeningeal angioma in a child where conventional post-gadolinium T1-weighted images failed to show clear leptomeningeal enhancement. Although more data are warranted, these reports indicate that MR venography can be a useful, complementary radiological tool to evaluate sinuses and large veins, and post-contrast TOF images can be very sensitive in detecting small angiomas otherwise missed by conventional sequences.

MRI Volumetry Studies. MRI volumetry of the brain, including both global volumes and segmented grey and white matter volumes, is an objective, quantitative imaging approach to assess structural integrity. Both grey and white matter volumes are related to intelligence in normal humans, including children,[50] and global or regional cerebral volume losses are also known to be associated with neuro-cognitive dysfunctions in various neurological disorders. In clinical practice, MR volumetry is most often used in measuring hippocampal volumes to detect hippocampal volume loss associated with temporal lobe epilepsy. Hemispheric volume measurements are currently not a part of the routine clinical MRI work-up in Sturge-Weber syndrome (or any other neurological disorders), but they can be used to study effects of brain tissue loss on various neuro-cognitive functions. In a study of 18 subjects (with ages ranging from 4 months through 35 years but with 15 children) with Sturge-Weber syndrome and unilateral hemispheric involvement, brain volumes were measured on T1- and T2-weighted images and hemispheric atrophy was characterized by laterality scores of hemispheric volumes.[51] The authors found a good correlation between hemispheric atrophy and overall clinical severity scores and also with hemiparesis subscores; correlation with cognitive scores was weaker. In a subsequent study of 21 children with unilateral Sturge-Weber syndrome, Juhász and associates[52] reported that segmented, hemispheric gray- and white matter volumes on the angioma side (but not on the contralateral side) are both strongly related to the IQ of these children, but only white matter volumes proved to be significant, independent predictors of cognitive functions in multivariate analyses. These results further underline the importance of white matter integrity during the course of the disease. Therefore, both water diffusion and MR volumetry studies suggest that protection of white matter integrity during the early phase of the disease (before irreversible atrophy occurs), would be a powerful approach to prevent poor neuro-cognitive outcome in Sturge-Weber syndrome. For example, since N-methyl-D-aspartate (NMDA) receptors expressed on oligodendrocytes play a crucial role in ischemia-induced white matter injury,[53] novel NMDA-antagonist drugs may target this mechanism and offset the consequences of white matter damage in affected children.

[^{1}H] Magnetic Resonance Spectroscopy (MRS). MR spectroscopy was one of the first MR applications where functional aspects of the brain tissue could be studied and even quantitatively evaluated through measuring concentrations of several metabolites. MRS uses frequency shifts of metabolites to create a profile within a given region of interest. Most clinical MR spectroscopy studies have been carried out using proton [^{1}H] MRS, although feasibility of ^{31}P, ^{13}C, ^{23}Na, and ^{18}F MRS has been also established for human use. [^{1}H]MRS allows quantitative *in vivo* measurements of a number of biochemicals, including N-acetyl-aspartate (NAA), choline, creatine/phosphocreatine (Cr/PCr), lactate, and glutamate/glutamine/gamma-aminobutyric acid (Glx) (these latter three compounds can only be separated using high-field magnets).[54] NAA is almost exclusively confined to neurons, which makes it a useful marker for neuronal injury. Cr/PCr can be found in both neurons and glia (although glia has a higher concentration of this compound), and complete absence of this signal points to neuronal loss. The choline peak is a combination of glycerolphosphocholine and phosphocholine, which are markers of membrane turnover and degradation; increased choline concentration is consistent with myelin breakdown in the brain. Glx concentration measured by MRS is likely related to the metabolic state of the tissue, reflecting a tight coupling between glutamate turnover and glucose metabolism in both normal and epileptic cerebral cortex.[55] Lactate is not found in healthy tissue; elevated lactate levels can reflect impaired oxidative respiration and use of carbohydrate catabolism. Lactate levels can rise due to a switch in metabolism from aerobic to anaerobic glycolysis, and this is frequently observed in the presence of malignant tumors.[56] Increased lactate concentrations can also be found in the ischemic penumbra and even on the contralateral side of large infarcts after ischemic stroke.[57]

Initial single-voxel MRS studies in children with Sturge-Weber syndrome showed a significant decrease of NAA concentrations in the gadolinium-enhancing volumes of interest, suggesting that MRS may be useful for early characterization and monitoring of neuronal dysfunction or loss in this disease (**Figure 6-9**).[58] More recently, chemical shift imaging (CSI) or magnetic resonance spectroscopic imaging (MRSI) techniques allowed measurement of MRS metabolites in multiple voxels (typically in one or more 1-2 cm thick transverse planes centered over presumably affected or other targeted brain regions) simultaneously. MRSI studies in Sturge-Weber syndrome demonstrated abnormal metabolite concentrations (decreased NAA, increased choline peaks) in not only posterior regions abnormal on conventional MRI, but also in structurally unaffected grey matter tissue in the frontal lobe.[36,59,60] Biochemical abnormalities can even be detected in areas showing no abnormalities on diffusion tensor imaging.[59] Importantly, NAA decreases were more severe in children with early seizure onset, and frontal lobe NAA asymmetry was also a good predictor of motor functions.[35,60]

Limitations of MRS/MRSI studies include limited spatial coverage (lack of true 3D-imaging, which can lead to sampling errors) and spatial resolution limited by voxel size as well as length of image acquisition. MR spectra are also highly sensitive for movement artifacts, although this is rarely a problem in sedated children. Clinical use of high-magnetic field scanners (3 Tesla and above) can partially overcome these limitations and also allow detection and quantification of a number of additional compounds to gain novel insights in biochemical abnormalities associated with various neurological disorders including Sturge-Weber syndrome.

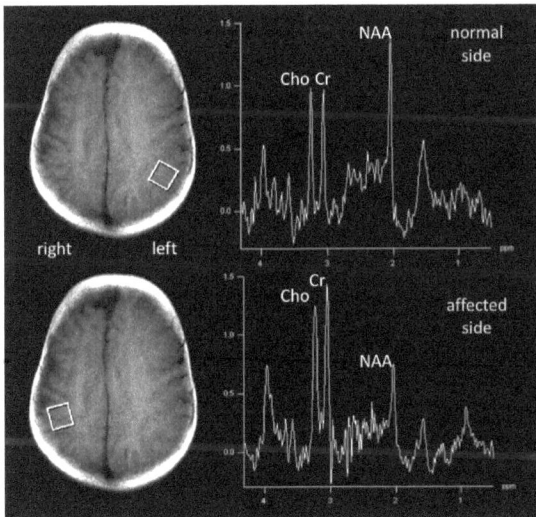

Figure 6-9. Single voxel [¹H] MR spectroscopy of a 10-month-old infant with a right facial port wine stain and seizures. Conventional MRI showed early signs of atrophy in the right hemisphere. Upper panel: Voxel location and spectrum from the left (unaffected) parietal lobe. Lower panel: Voxel location and spectrum from the right (affected) parietal lobe. Note the considerably decreased NAA peak and slight increase of the choline (Cho) (and creatine [Cr]) peaks, resulting in a markedly decreased NAA/Cho ratio. Subsequent MRIs demonstrated progressive atrophy of this brain region.

Functional MRI (fMRI) Techniques. FMRI measures the hemodynamic response (characterized by a dynamic change in blood flow) related to neural activity in the brain; thus, fMRI can be considered as a typical functional imaging modality, which has largely replaced other riskier (due to radiation exposure) and more expensive imaging modalities (such as blood flow SPECT or PET studies) in the study of brain activation. The technique is based on measurement of the blood-oxygen-level dependent (BOLD) signal. The measured functional contrast is attributed to the iron in hemoglobin that becomes paramagnetic when it is deoxygenated, thus producing a local susceptibility increase. This change in hemoglobin oxygenation can be observed using a variety of pulse sequences.[61] A typical BOLD response represents a relatively small (1% to 5%) change in regional image intensity, which develops over 3 to 8 seconds following task initiation. Task-specific BOLD signal changes are not directly quantifiable but are expressed as a percentage signal change or as a statistical significance level based on a particular statistical model.

Functional MRI is an increasingly utilized technique in the presurgical evaluation of epilepsy, in both adults and children.[62] A common clinical application of fMRI is non-invasive mapping of eloquent cortex (mostly motor and language areas) in patients with lesions, including children, to assess resectability and reduce the possibility of functional deficit.[63] This can be relevant in patients with Sturge-Weber syndrome associated with medically refractory epilepsy, if partial resection, rather than hemispherectomy, is considered. In these cases, preservation of the sensori-motor cortex, often located adjacent to (commonly in front of) the angioma, is a common issue. Despite some physiologic differences, the basic BOLD response in children is similar to that of adults, with some task-related differences.[64] However, since activation fMRI techniques require patient co-operation, only children five years of age or above (also depending on their cognitive function) can be reasonably tested by this approach;[62, 65, 66] older children and girls have a higher success rate to complete the tasks.

In children with Sturge-Weber syndrome and epilepsy affecting the left hemisphere, fMRI can be clinically useful to evaluate potential reorganization of language functions in the contralateral hemisphere. This is a clinically important issue because previous activation studies using ^{15}O-PET have demonstrated that, in some cases, language processing continues to involve hypoperfused, structurally damaged, and even calcified areas in Sturge-Weber syndrome.[67] Therefore, demonstration of at least partial reorganization of language functions in the right hemisphere can increase the confidence of good post-surgical language outcome in patients where hemispherectomy is planned. This is particularly critical in older children or adults where plasticity of language (and other) functions becomes increasingly limited.

Other functions such as visual or auditory processing can also be evaluated by fMRI. Visual and auditory cortices can be activated even in sedated children. Normal subjects show bilateral activation in primary visual cortex (Brodmann's areas 17 and 18), sometimes extending to secondary visual areas.[68] Patients with Sturge-Weber syndrome can demonstrate variable activations in the affected occipital lobe, including increased activation, no activation, or abnormal distribution of the activation.[69] These findings again demonstrate that presence of the vascular malformation does not necessarily prevent cortical activation in the occipital cortex. Thus, assessing cortical function by fMRI in patients with Sturge-Weber syndrome may be helpful in surgical decisions.

Functional reorganization can be studied by fMRI after resective surgery. This has been demonstrated by an MRI study of a girl who underwent a left occipitotemporal resection due to Sturge-Weber syndrome associated with seizures at five years of age.[70] Functional MRI with activations of both visual word and speech processing at demonstrated a selective shift of the visual component of reading to the right hemisphere, whereas the verbal components remained in the left hemisphere. One could hypothesize that there was a selective and successful shift to the right hemisphere of the visual component of reading because the lesion occurred before the acquisition of literacy, while verbal transcoding and output remained in the left hemisphere. Importantly, demonstration of the absence of fMRI activation in the residual left ventral temporal cortex contributed to the decision to perform a further occipitotemporal resection due to seizure recurrence, without any functional sequelae.

It should be noted that the fMRI technique may also be used for interictal localization of epileptic cortex, through detecting cerebral hemodynamic changes produced by epileptiform discharges. EEG-triggered fMRI, with a high spatial and temporal resolution, can assist detection of the spatiotemporal pattern of spike origin and propagation.[71] However, the technique has inherent limitations arising from the limitations of detecting interictal spikes on scalp EEG and other technical issues.

Finally, resting-state fMRI represents a relatively novel and different paradigm where no activations are obtained and analyzed; rather, functional connectivity of different brain regions can be assessed by various algorithms calculated from MR signals recorded during baseline brain activity (i.e., during rest, representing the "default mode" network). This technique has several advantages over activation fMRI studies. For example, data acquisition requires less co-operation during resting-state than with predefined active tasks, which can be advantageous when studying children. Resting-state connectivity changes have been shown to correlate with various cognitive and behavioral characteristics,[72,73] and altered functional connectivity was also found in the epileptic brain.[74] This technique holds the promise of revealing abnormalities of brain network connectivity in neuropsychiatric disorders. These abnormalities could become early, objective imaging markers of various behavioral and other conditions. This is a rapidly evolving field of research, with efforts to overcome several pitfalls of the method.[75]

Imaging Ocular Abnormalities. CT scan of the ocular bulb in Sturge-Weber syndrome may show globe enlargement (buphthalmos) and ocular enhancement due to choroidal disease,[76] while MR imaging can demonstrate choroidal hemangioma. However, ocular abnormalities do not correlate with the extent of hemispheric involvement.[77] In about half of the cases, the ocular globe wall is thickened posteriorly, and thinned as it extends to the ciliary bodies. The low-flow vascular malformation may result in choroidal or retinal detachment, and this, as well as globe enlargement and hemorrhage can be visualized by orbital MR imaging.

PET and SPECT Imaging

Both positron emission tomography (PET) and single photon emission computed tomography (SPECT) are non-invasive functional neuroimaging modalities that depict local chemical functions in various body organs. The PET technique employs a camera with multiple pairs of oppositely situated detectors, which are used to record the paired high-energy (511 KeV) photons traveling in opposite directions as a result of positron decay. In general, PET images can reach an excellent spatial resolution in the mm range. Beginning in the early 2000's, traditional PET cameras have been gradually replaced by PET/CT scanners, which use low-energy CT-scans to achieve attenuation correction and, thus, decrease total scanning times. New generation MR-PET cameras that are being developed will provide a new level of multi-modal imaging showing metabolic abnormalities with superior anatomical detail and direct comparison of functional and anatomic abnormalities.

The most widely used PET tracer is 2-deoxy-2[^{18}F]fluoro-D-glucose (FDG) to measure glucose metabolism. Under steady-state conditions, FDG uptake reflects the utilization rate of glucose by an organ. In the brain, this rate is among the highest in the body, and its value is highly related to the synaptic density and functional activity of the brain tissue. FDG uptake is generally decreased in sites with neuronal loss and in epileptic foci during the interictal state. However, clinical or electrographic seizures and even frequent interictal spiking can increase glucose metabolism. Therefore, FDG PET studies in patients with epilepsy should be done in tandem with scalp EEG monitoring to avoid false interpretation of ictal/post-ictal images or interictal hypermetabolism caused by frequent spiking.

When interpreting brain FDG PET images in children, one has to take the age of the patient into account. The pattern of glucose utilization undergoes dramatic changes in the first postnatal year. In the newborn, the highest glucose metabolic activity is seen in primary sensorimotor cortex, thalamus, brainstem, and cerebellar vermis.[78-80] Subsequently, the ontogeny of regional brain glucose metabolism follows a phylogenetic order, with functional maturation of older anatomical structures preceding that of newer areas. Although there are large nonlinear changes in the absolute values of regional brain glucose metabolic rates between one year of age and adulthood, the overall pattern of brain glucose metabolism at one year of age appears similar to that seen in adults. Thus, in those pediatric neurological disorders, which basically affect only one hemisphere (such as most cases of Sturge-Weber syndrome), the contralateral hemisphere can be used as an internal control for assessing focal metabolic abnormalities at different stages of brain maturation in early childhood.

In comparison to PET, SPECT has lower spatial resolution (in the cm range) and this technique is not able to provide absolute quantification; rather, quantitative estimations in brain SPECT studies are performed by calculating tracer uptake ratios. On the other hand, SPECT instrumentation is far less costly as it does not require operation of a cyclotron, and radiolabeled tracers are commercially available. The most common clinical use of SPECT is the study of altered regional cerebral blood flow for which the most frequently used tracers are ^{99m}Tc-hexamethylpropyleneamine-oxime (^{99m}Tc-HMPAO) and ^{99m}Tc-ethyl cysteinate dimer (^{99m}Tc-ECD). These tracers have a quick initial uptake in brain and reach the peak within two minutes of injection, without redistribution. Thus, the initial tracer uptake and distribution correspond to regional blood flow at the time of injection and remain unchanged for hours, independent of blood flow variations occurring after the fixation time. ^{99m}Tc-ECD has some advantages against ^{99m}Tc-HMPAO, such as better *in vitro* stability and rapid clearance from extracerebral tissue resulting in better brain-to-background ratio and superior image quality. Nevertheless, both these radiotracers can be injected into the patient during an epileptic seizure and the actual image acquisition can be performed later after recovery from the seizure. For imaging (lateralization and localization of) epileptic foci, ictal SPECT is superior to interictal SPECT, as demonstrated by several studies of patients with temporal and extratemporal lobe epilepsy (reviewed by la Fougere and colleagues[81]); however, timing of injection is critical to obtain optimal results. Late (after >30 sec following the earliest signs of seizure onset) tracer injection may result in difficult or even false interpretations of the obtained blood flow patterns as regions with rapid seizure spread may show more increases than seizure onset areas. Also, ictal SPECT imaging of seizure foci associated with very brief seizures (e.g., infantile spasms or myoclonic seizures) is difficult and often not feasible clinically. Currently, the most accurate focus localization results can be achieved by applying subtraction of interictal from ictal SPECT scans, and co-registration of the subtraction images with the patient's own high-resolution MR; this technique is commonly called SISCOM (subtraction ictal SPECT co-registered to MRI[82]).

PET and SPECT imaging are currently not part of the routine radiological evaluation of Sturge-Weber syndrome; rather, they are mostly reserved for selected patients where seizures are uncontrolled and surgical resection is contemplated. However, we find that FDG PET studies are useful in evaluating brain dysfunction associated with cognitive and neurological impairment and their progression in affected children. Indeed, clinical use of PET imaging may be expanding beyond epileptic focus localization as our knowledge regarding metabolic correlates of neuro-cognitive functions deepens.

PET in Sturge-Weber Syndrome. Initial studies with FDG PET in Sturge-Weber syndrome demonstrated reduced glucose metabolism of the cortex underlying the leptomeningeal lesion, but also almost invariably extending beyond the area of vascular malformation depicted by structural imaging (**Figure 6-10**).[83] The most severe hypometabolism was seen in calcified cortex. Subsequent PET studies, measuring hemispheric extent of cortical hypometabolism in children with unilateral Sturge-Weber syndrome, confirmed these results and also showed that relatively mildly hypometabolic cortex (10% to 20% decrease as compared to the contralateral normal cortex metabolism) was more extensive in patients with frequent seizures.[84] This supported the previously postulated notion that frequent (intractable) seizures, in addition to hypoxic injury due to venous congestion, may play a role in progression of cortical hypometabolism beyond obvious structural damage. This notion has been further corroborated by a longitudinal PET study in children with Sturge-

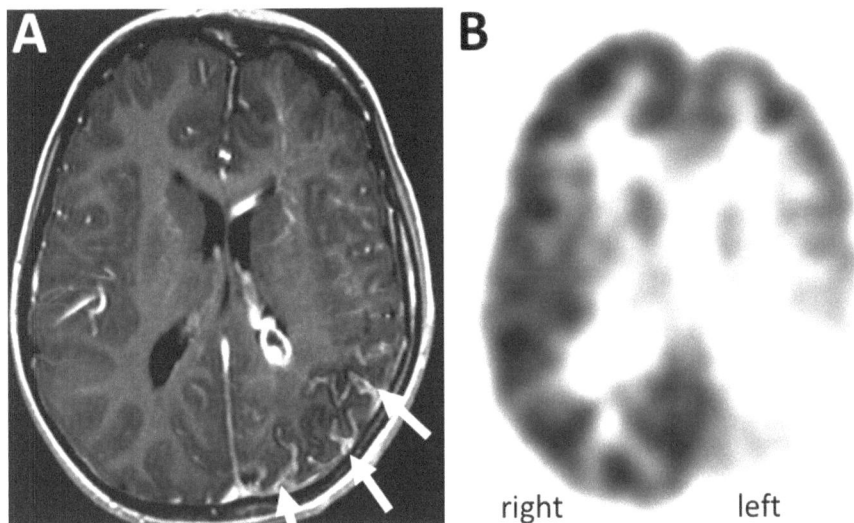

Figure 6-10. Comparison between abnormalities seen on T1-weighted, post-gadolinium MRI (A) and glucose (FDG) PET (B) images in a patient with Sturge-Weber syndrome and left hemispheric involvement. MRI showed leptomeningeal enhancement in the posterior cortex (arrows), along with enlarged choroid plexus in the posterior horn of the left lateral ventricle. Some deep draining veins are also seen in the prefrontal region. FDG PET showed extensive hypometabolism, with most severe involvement in the posterior cortex underlying the angioma. This posterior cortex also showed calcification. The entire frontal lobe was also moderately hypometabolic.

Weber syndrome.[85] In this study of 14 children, all below four years of age at the time of the initial scans, the authors found robust interval increase in the extent of hypometabolic cortex (from an average of 17% of the hemispheric surface to 58%) in six subjects (**Figure 6-11**); this major expansion of hypometabolism always occurred before three years of age, a period when the maturing normal brain undergoes a massive, developmentally regulated increase in brain glucose metabolism. Importantly, these children with Sturge-Weber syndrome had higher clinical seizure frequency during the follow-up period than those with no signs of metabolic progression. Also, somewhat unexpectedly, a few children showed an interval decrease in the size of hypometabolic cortex. Seizures in these latter children became relatively well controlled during the months before the second scan. Reversibility of cortical hypometabolism, related to seizure control, had been shown in previous longitudinal PET studies in children with epilepsy (not related to Sturge-Weber syndrome).[86] Adaptive mechanisms,

FDG PET 1   FDG PET 2

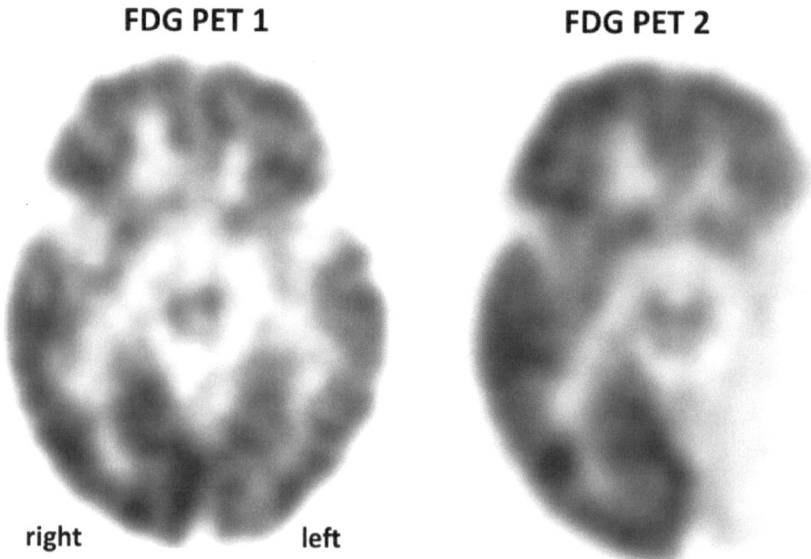

right   left

Figure 6-11. Rapid progression of a left temporo-occipital glucose hypometabolism in an infant with Sturge-Weber syndrome. Initial FDG PET scan at 8 months of age showed mildly decreased glucose utilization in this region. A repeated PET scan seven months later, however, demonstrated an almost complete loss of metabolic activity of the same area. The infant had rare seizures.

e.g., development of effective collateral circulation, could also play a role in this metabolic improvement and contribute to the observed improvement of hemiparesis in the same children.

Altogether, these PET findings strongly suggest that chronic seizure activity plays a role in development and progression of cortical metabolic abnormalities in affected children. Indeed, cerebral blood flow studies during seizures can demonstrate a steal phenomenon, leading to ischemic conditions in primarily unaffected brain regions, at the time when seizure activity is present.[87] These metabolic imaging data provide support for the argument for early, aggressive seizure control in children diagnosed with Sturge-Weber syndrome.[88]

In addition to hypometabolism, a unique, paradoxical pattern of interictal glucose *hyper*metabolism was observed on FDG PET in some infants scanned before one year of age (**Figure 6-12**).[83] In a subsequent study, this metabolic pattern was observed in almost half of children with Sturge-Weber syndrome who underwent PET scanning before 2 years of age, including two children who were scanned before the onset of their first clinical seizures.[89] EEG monitoring during the tracer uptake period showed no signs of seizure activity in any of the affected patients; therefore, subclinical seizures or frequent interictal epileptiform activity could not explain increased metabolism in these cases. Alternatively, ischemic neuronal damage causing astroglial activation could account for the increased glucose utilization.[90,91] Activated astrocytes may contribute to a decline of neurologic function by release of glutamate and other excitotoxic amino acids.[92] Similar metabolic phenomena have been observed in neonates following perinatal hypoxia. In those cases, transient glucose hypermetabolism can be observed in the basal ganglia, a structure metabolically active at the time of the early ischemic event.[93] Increased glutamate concentrations in the affected basal ganglia, measured by MRS, provided imaging evidence of abnormal glutamatergic mechanism in post-hypoxic brain tissue abnormalities.[94] Because glutamate is an excitatory neurotransmitter, an imbalance between glutamate

1 .9 years old   2.7 years old

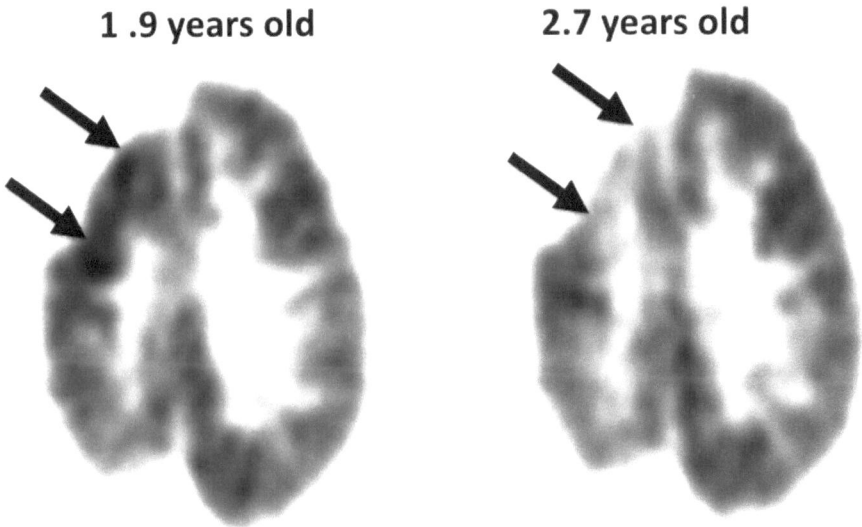

Figure 6-12. FDG PET demonstrating a transient increase of interictal glucose metabolism in a young boy with Sturge-Weber syndrome and a right hemispheric angioma. In this case increased glucose uptake was seen on the first scan, shortly before the first seizures developed. EEG monitoring during the tracer uptake period showed no epileptiform activity (electrographic seizures or spiking), indicating the true interictal nature of this metabolic abnormality. A second scan eight months later showed hypometabolism in the same region. The patient developed intractable epilepsy between the two scans.

and inhibitory neurotransmitters (particularly GABA) may also contribute to epileptogenicity. Whether increased glucose metabolism is a poor prognostic sign for uncontrolled seizures in infants with Sturge-Weber syndrome remains to be clarified. In the only series where this was studied retrospectively, four of eight patients showing this early metabolic pattern developed uncontrolled seizures requiring hemispherectomy.[89] Importantly, in all but one case, where a follow-up PET scan was performed, hypermetabolic regions became hypometabolic initially, suggesting that cortical areas showing early hypermetabolism are at high risk for subsequent tissue damage. Similar findings of brain perfusion have been reported in SPECT studies (see SPECT section below).[95] Altogether, focal hypermetabolism in young children with Sturge-Weber syndrome may reflect a transient increase of metabolic demand in cortex undergoing excitotoxic tissue damage. This period likely indicates a critical time window when therapeutic interventions (e.g., emerging neuroprotective or antiepileptogenic treatment) could be most effective in preventing severe epilepsy and diminishing long-term tissue damage.

The severity and extent of cortical and subcortical glucose hypometabolism, measured by PET, are important imaging markers of cognitive functions and may have predictive values for cognitive outcome in children with Sturge-Weber syndrome. The relationship between cortical hypometabolism and cognitive functions in patients with unilateral hemispheric involvement is complex, probably influenced by the timing of early unilateral progression (leading to impaired function of the affected brain areas), which may facilitate effective reorganizational processes in unaffected regions of the developing brain. An initial evidence for such a process was provided by the studies of Lee and colleagues[84] demonstrated a paradoxical preservation of IQ in some children who underwent an early, rapid hemispheric progression reflected by severe, extensive hypometabolism on FDG PET in the affected hemisphere. It is likely that early unilateral progression in these cases prompted the opposite hemisphere (not involved in the pathology) to take over critical cognitive functions resulting in a surprisingly good outcome (**Figure 6-13A**). On the other hand, patients with smaller areas of hypometabolism, where progression is slow, often show unexpectedly poor cognitive outcome, especially when seizures cannot be controlled (**Figure 6-13B**). The findings of multimodal imaging studies (comparing MRI and PET images) strongly suggest that ipsilateral cortical areas that appear to be structurally intact (normal on MRI) but function at a suboptimal level (manifested as mild hypometabolism on FDG PET) are associated with adverse cognitive outcome, possibly because of inadequate

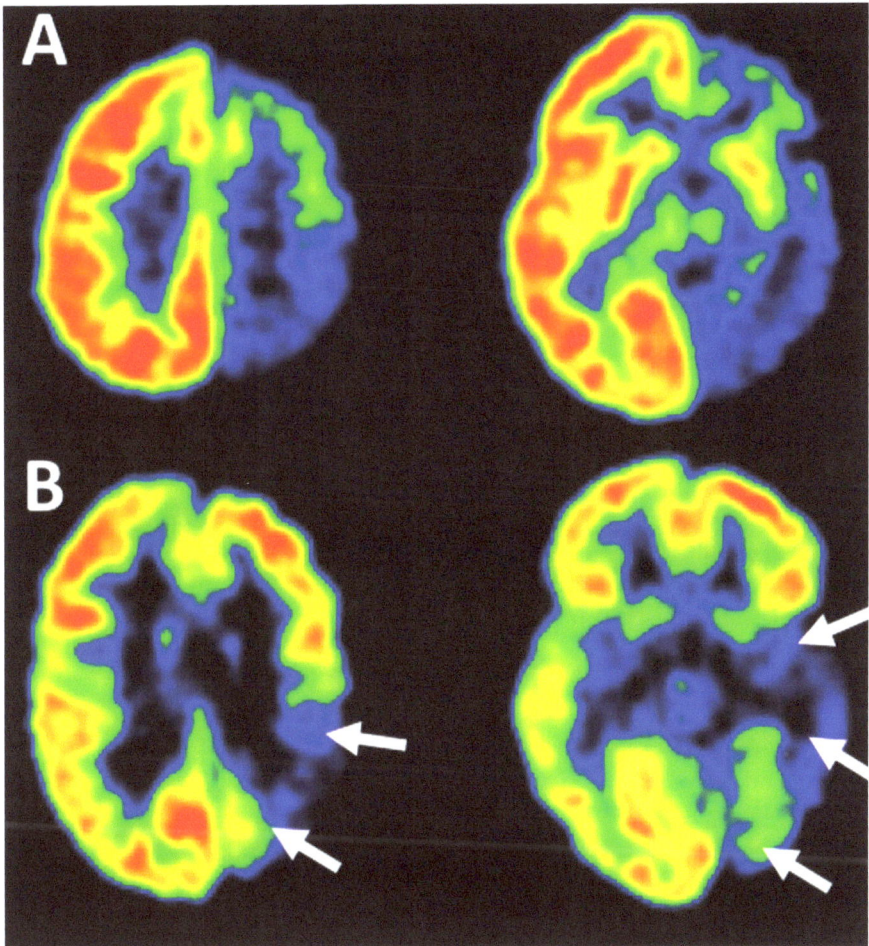

Figure 6-13. FDG PET images of two children with Sturge-Weber syndrome and left hemispheric involvement. Both patients had occasional seizures. (A) This child had an early, rapid hemispheric progression of hypometabolism (before 3 years of age), with relatively well-preserved cognitive functions (IQ = 79, good verbal skills). (B) This child, with a left temporo-parieto-occipital hypometabolism, had a full-scale IQ of 55. These two cases demonstrate a paradoxical relation between extent of cortical hypometabolism and cognitive functions: early, extensive hemispheric progression is often associated with relatively good cognitive outcome, presumably because of effective reorganization processes in the intact hemisphere.

reorganizational processes in other brain regions. Effective reorganization may also not occur because repeated seizures from the affected hemisphere exert a negative influence on the contralateral hemisphere. This is consistent with suggestions that the earlier the "good" hemisphere is spared from the detrimental effect of sustained epileptic discharges originating in the affected hemisphere, the more effectively sensorimotor, intellectual, and psychosocial development can take place in the remainder of the brain.[96] Thus, surgical elimination of the hypometabolic cortex may prevent further cognitive decline (and even facilitate cognitive improvement) even if seizures are not devastating. Repeated PET scanning may be helpful to track metabolic progression and optimize the timing of surgery.

In addition to cortical metabolic abnormalities, thalamic hypometabolism on the side of the angioma is also often seen in children with Sturge-Weber syndrome. Thalamic metabolic asymmetry is easy to measure on PET and can be a simple functional imaging marker of cognitive functions. Indeed, a strong correlation was found between thalamic metabolic asymmetries and full-scale IQ in 20 children (age: 3-12 years) with unilateral Sturge-Weber syndrome.[42] Severe thalamic asymmetries of glucose uptake and also diffusivity (measured by DTI) were strong predictors of full-scale IQ, even after controlling for age and extent of cortical hypometabolism measured in the same hemisphere. In addition, thalamic glucose hypometabolism was also associated with impairment of motor functions. This is not surprising, as the thalamus is a central relay station and integrator of sensorimotor functions, and this deep brain structure could be affected both directly (e.g., by venous congestion in the deep veins) and indirectly (via cortico-thalamic connections) in Sturge-Weber syndrome.

PET studies can be informative not only in revealing the extent of functional brain damage but also in highlighting preserved functions and functional reorganization. In unilateral Sturge-Weber syndrome, there is a great potential for interhemispheric reorganization of critical brain functions because the contralateral hemisphere is basically normal early in the course of the disease.[97] For example, studies using [^{15}O]water PET, an imaging approach able to visualize brain regions with significant blood flow increases during activations by various (e.g., motor or language) tasks, showed that interhemispheric reorganization is more pronounced for language than for motor functions in unilateral Sturge-Weber syndrome.[67] This activation pattern was similar to that in patients with early left hemispheric lesion from other causes.[98] The same studies have also demonstrated that hypoperfused posterior brain regions with apparently severe structural damage, such as calcification, can become activated during language and motor performance. Thus, functional imaging can be helpful to carefully mapping eloquent cortex, which is of vital importance in the case of presurgical evaluation, especially when partial resection is planned. Due to its lack of radiation exposure and wider availability, functional MRI (see above) has largely replaced activation PET studies for presurgical brain mapping, and such activation PET studies are largely reserved for patients for whom MRI scanning is contraindicated (e.g., because of presence of a metal device in the body).

Evidence supporting functional reorganization in the hemisphere contralateral to the angioma was provided by an interictal FDG PET study.[99] In a group of 17 children (aged 1.7 to 10.3 years), increased glucose utilization was found in the occipital (primary visual) and parietal cortex contralateral to the angioma and hypometabolic parieto-occipital regions. FDG uptake was especially high in contralateral visual cortex of children showing severe occipital hypometabolism on the angioma side (**Figure 6-14**). This may be due to increased synaptic density in this area, presumably from compensatory processes induced by early, severe injury in the affected visual cortex. Mechanisms and clinical relevance of these metabolic increases remain to be elucidated.

SPECT Studies. There have been several SPECT studies evaluating regional cerebral perfusion in patients with Sturge-Weber syndrome. As expected, decreased perfusion on interictal SPECT is commonly seen in brain regions showing structural abnormalities on CT or MRI scans.[100-102] In addition, similar to FDG PET abnormalities, perfusion deficits measured by SPECT

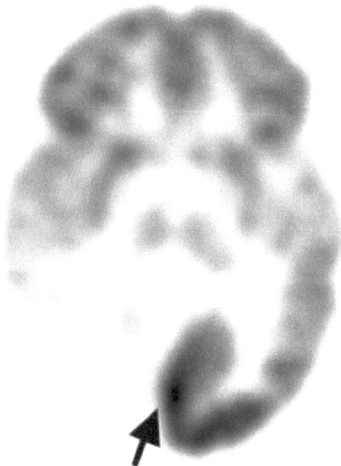

Figure 6-14. Increased glucose metabolism (arrow) in the left occipital (visual) cortex, contralateral to the leptomeningeal angioma, in a child with unilateral Sturge-Weber syndrome. His right occipital lobe became severely atrophic and hypometabolic early in the course of the disease.

often extend beyond apparent structural lesions,[100,103,104] although this is not always the case. Griffiths and colleagues[101] reported that the area of abnormal perfusion, measured by HMPAO SPECT, was actually smaller than the extent of leptomeningeal enhancement in about half of their patients (all were children).

During the course of the disease, progressive hypoperfusion measured with SPECT were associated generally with neurologic deterioration.[105] Early metabolic dysfunction has been described in some children with port-wine stains and neonatal seizures, even before clear signs of atrophy or calcification.[102] With advances in MRI technology, early perfusion deficits can be detected by perfusion MRI (see above) performed during the clinical MRI protocol; therefore interictal SPECT imaging is generally not warranted in young children where intracranial involvement of Sturge-Weber syndrome is to be determined.

Nevertheless, an important contribution to the understanding of the pathophysiology of Sturge-Weber syndrome came from longitudinal SPECT studies done by Pinton and associates[95] in 22 infants (all of whom were less than 25 months old at the time of enrollment). Twelve of the patients had no seizures before the initial SPECT studies (all of them were less than four months old at the first scan), and seven of them underwent a follow-up SPECT four to 11 months later. Interestingly, while all children with seizures showed impaired interictal perfusion in the affected hemisphere, nine of the 12 infants with no seizures initially had positive hemispheric indices, indicating increased perfusion ipsilateral to the angioma. Moreover, the second SPECT scans demonstrated an interval switch from hyper- to hypoperfusion in the seven patients where such follow-up scans were done, including in three children who had onset of seizures between the two studies.

These early perfusion findings, showing a transient increase of blood flow in the interictal state, were similar to those reported in infants who underwent interictal FDG PET (see PET section). Both perfusion and glucose metabolic studies consistently demonstrated that these functional abnormalities may be present before any neuro-cognitive symptoms occur, and the affected brain tissue is at high risk for subsequent functional impairment. Increased perfusion can coincide with accelerated myelination (see more details in MRI section), and excess perfusion likely plays a role in development of this white matter phenomenon observed on MRI in young infants affected by Sturge-Weber syndrome.[25] It is still questionable whether every patient undergoes this hyperperfusion/hypermetabolic phase, or if this represents one of several pathologic mechanisms. Performing functional imaging studies in the asymptomatic phase of the disease may become important if novel treatments targeting epileptogenesis or hypoxic brain injury become available, and selection of patients early in the course is desired.

Ictal SPECT studies are able to evaluate the hemodynamic response to seizures in patients with Sturge-Weber syndrome. Aylett and colleagues[106] combined SPECT with transcranial Doppler and found hemispheric hypoperfusion interictally in all three children who underwent these studies; ictal hyperperfusion was associated with increased blood flow velocity in the middle cerebral arteries; this was observed in both the affected (seizing) and contralateral hemisphere. Repeated recordings in one infant showed a diminishing hemodynamic response to seizures of the unaffected hemisphere. The results demonstrated that the venous malformation in Sturge-Weber syndrome is associated with an impairment of the cerebral hemodynamic response to seizure activity. Impaired vasoreactivity of cortical regions outside the structural lesion in Sturge-Weber syndrome patients with progressive disease[107,108] suggests that functional brain abnormalities extending beyond the structural lesion are important determinants of neurological and behavioral progression and should be considered when assessing therapeutic possibilities. In addition, motor function impairment correlates with regional cerebral blood flow better than with structural abnormalities seen on conventional imaging.[109] Ictal SPECT studies can be also useful to study residual epileptic foci or propagation patterns following functional hemispherectomy if the patient continues to have seizures.[110]

Multimodality Imaging

As the previous sections demonstrated, various structural and functional imaging modalities can depict a wide range of abnormalities in Sturge-Weber syndrome. MRI is generally a more sensitive means to visualize in venous abnormalities and white matter pathology, while PET and SPECT scans are more accurate in delineating the true extent and severity of functional abnormalities in the cortex and deep grey matter. Therefore these various imaging modalities provide complementary information, and combination of these, i.e., the use of multimodality imaging, can yield optimal radiological information regarding the severity and extent of brain involvement in this disorder. As discussed above, several imaging markers have emerged as important correlates of various neurological and cognitive functions in children with Sturge-Weber syndrome. Quantitative imaging variables, derived from multimodality imaging, play an increasingly important role in detecting early brain abnormalities. Such quantitative imaging markers can be informative for the projected course of the disease and potentially allow early, aggressive intervention to prevent irreversible neurocognitive sequelae of the disease. These imaging markers will be also instrumental to monitor effects of new treatment modalities and provide an objective measure of functional reorganization. Thus, advanced structural and functional imaging is expected to play an increasing role in the diagnosis and management of Sturge-Weber syndrome.

REFERENCES

1. Dimitri V. Tumor cerebral congenito (angioma cavernoso). *Rev Assoc Med Argent.* 1923;36: 1029-1037.

2. Poser CM, Taveras JM. Cerebral angiography in encephalo-trigeminal angiomatosis. *Radiology.* 1957;68(3): 327-336.

3. Bentson JR, Wilson GH, Newton TH. Cerebral venous drainage pattern of the Sturge-Weber syndrome. *Radiology.* 1971;101(1): 111-118.

4. Kuhl DE, Bevilacqua JE, Mishkin MM, Sanders TP. The brain scan in Sturge-Weber syndrome. *Radiology.* 1972;103(3): 621-626.

5. Welch K, Naheedy MH, Abroms IF, Strand RD. Computed tomography of Sturge-Weber syndrome in infants. *J Comput Assist Tomogr.* 1980;4(1): 33-36.

6. Gobbi G, Ambrosetto P, Zaniboni MG, Lambertini A, Ambrosioni G, Tassinari CA. Celiac disease, posterior cerebral calcifications and epilepsy. *Brain Dev.* 1992;14(1): 23-29.

7. Wasenko JJ, Rosenbloom SA, Duchesneau PM, Lanzieri CF, Weinstein MA. The Sturge-Weber syndrome: comparison of MR and CT characteristics. *AJNR Am J Neuroradiol.* 1990;11(1): 131-134.

8. Marti-Bonmati L, Menor F, Poyatos C, Cortina H. Diagnosis of Sturge-Weber syndrome: comparison of the efficacy of CT and MR imaging in 14 cases. *AJR Am J Roentgenol.* 1992;158(4): 867-871.

9. Aguglia U, Latella MA, Cafarelli F, et al. Spontaneous obliteration of MRI-silent cerebral angiomatosis revealed by CT angiography in a patient with Sturge-Weber syndrome. *J Neurol Sci.* 2008;264(1-2): 168-172.

10. Elster AD, Chen MY. MR imaging of Sturge-Weber syndrome: role of gadopentetate dimeglumine and gradient-echo techniques. *AJNR Am J Neuroradiol.* 1990;11(4): 685-689.

11. Sperner J, Schmauser I, Bittner R, et al. MR-imaging findings in children with Sturge-Weber syndrome. *Neuropediatrics.* 1990;21(3): 146-152.

12. Lipski S, Brunelle F, Aicardi J, Hirsch JF, Lallemand D. Gd-DOTA-enhanced MR imaging in two cases of Sturge-Weber syndrome. *AJNR Am J Neuroradiol.* 1990;11(4): 690-692.

13. Benedikt RA, Brown DC, Walker R, Ghaed VN, Mitchell M, Geyer CA. Sturge-Weber syndrome: cranial MR imaging with Gd-DTPA. *AJNR Am J Neuroradiol.* 1993;14(2): 409-415.

14. Adams ME, Aylett SE, Squier W, Chong W. A spectrum of unusual neuroimaging findings in patients with suspected Sturge-Weber syndrome. *AJNR Am J Neuroradiol.* 2009;30(2): 276-281.

15. Slasky SE, Shinnar S, Bello JA. Sturge-Weber syndrome: deep venous occlusion and the radiologic spectrum. *Pediatr Neurol.* 2006;35(5): 343-347.

16. Andeweg J. The anatomy of collateral venous flow from the brain and its value in aetiological interpretation of intracranial pathology. *Neuroradiology.* 1996;38(7): 621-628.

17. Andeweg J. Consequences of the anatomy of deep venous outflow from the brain. *Neuroradiology.* 1999;41(4): 233-241.

18. Griffiths PD, Blaser S, Boodram MB, Armstrong D, Harwood-Nash D. Choroid plexus size in young children with Sturge-Weber syndrome. *AJNR Am J Neuroradiol.* 1996;17(1): 175-180.

19. Mentzel HJ, Dieckmann A, Fitzek C, Brandl U, Reichenbach JR, Kaiser WA. Early diagnosis of cerebral involvement in Sturge-Weber syndrome using high-resolution BOLD MR venography. *Pediatr Radiol.* 2005;35(1): 85-90.

20. Griffiths PD, Coley SC, Romanowski CA, Hodgson T, Wilkinson ID. Contrast-enhanced fluid-attenuated inversion recovery imaging for leptomeningeal disease in children. *AJNR Am J Neuroradiol.* 2003;24(4): 719-723.

21. Shin RK, Moonis G, Imbesi SG. Transient focal leptomeningeal enhancement in Sturge-Weber syndrome. *J Neuroimaging.* 2002;12(3): 270-272.

22. Terdjman P, Aicardi J, Sainte-Rose C, Brunelle F. Neuroradiological findings in Sturge-Weber syndrome (SWS) and isolated pial angiomatosis. *Neuropediatrics.* 1991;22(3): 115-120.

23. Atlas SW, Grossman RI, Hackney DB, et al. Calcified intracranial lesions: detection with gradient-echo-acquisition rapid MR imaging. *AJR Am J Roentgenol.* 1988;150(6): 1383-1389.

24. Jacoby CG, Yuh WT, Afifi AK, Bell WE, Schelper RL, Sato Y. Accelerated myelination in early Sturge-Weber syndrome demonstrated by MR imaging. *J Comput Assist Tomogr.* 1987;11(2): 226-231.

25. Adamsbaum C, Pinton F, Rolland Y, Chiron C, Dulac O, Kalifa G. Accelerated myelination in early Sturge-Weber syndrome: MRI-SPECT correlations. *Pediatr Radiol.* 1996;26(11): 759-762.

26. Porto L, Kieslich M, Yan B, Zanella FE, Lanfermann H. Accelerated myelination associated with venous congestion. *Eur Radiol.* 2006;16(4): 922-926.

27. Maton B, Krsek P, Jayakar P, et al. Medically intractable epilepsy in Sturge-Weber syndrome is associated with cortical malformation: implications for surgical therapy. *Epilepsia.* 2010;51(2): 257-267.

28. Reichenbach JR, Venkatesan R, Schillinger DJ, Kido DK, Haacke EM. Small vessels in the human brain: MR venography with deoxyhemoglobin as an intrinsic contrast agent. *Radiology.* 1997;204(1): 272-277.

29. Reichenbach JR, Haacke EM. High-resolution BOLD venographic imaging: a window into brain function. *NMR Biomed.* 2001;14(7-8): 453-467.

30. Sehgal V, Delproposto Z, Haacke EM, et al. Clinical applications of neuroimaging with susceptibility-weighted imaging. *J Magn Reson Imaging.* 2005;22(4): 439-450.

31. Juhász C, Haacke EM, Hu J, et al. Multimodality imaging of cortical and white matter abnormalities in Sturge-Weber syndrome. *AJNR Am J Neuroradiol.* 2007;28(5): 900-906.

32. Hu J, Yu Y, Juhász C, et al. MR susceptibility weighted imaging (SWI) complements conventional contrast enhanced T1 weighted MRI in characterizing brain abnormalities of Sturge-Weber Syndrome. *J Magn Reson Imaging.* 2008;28(2): 300-307.

33. Wu O, Ostergaard L, Sorensen AG. Technical aspects of perfusion-weighted imaging. *Neuroimaging Clin N Am.* 2005;15(3): 623-637.

34. Evans AL, Widjaja E, Connolly DJ, Griffiths PD. Cerebral perfusion abnormalities in children with Sturge-Weber syndrome shown by dynamic contrast bolus magnetic resonance perfusion imaging. *Pediatrics.* 2006;117(6): 2119-2125.

35. Lin DD, Barker PB, Hatfield LA, Comi AM. Dynamic MR perfusion and proton MR spectroscopic imaging in Sturge-Weber syndrome: correlation with neurological symptoms. *J Magn Reson Imaging.* 2006;24(2): 274-281.

36. Lin DD, Barker PB, Kraut MA, Comi A. Early characteristics of Sturge-Weber syndrome shown by perfusion MR imaging and proton MR spectroscopic imaging. *AJNR Am J Neuroradiol.* 2003;24(9): 1912-1915.

37. Oguz KK, Senturk S, Ozturk A, Anlar B, Topcu M, Cila A. Impact of recent seizures on cerebral blood flow in patients with sturge-weber syndrome: study of 2 cases. *J Child Neurol.* 2007;22(5): 617-620.

38. Arulrajah S, Ertan G, A MC, Tekes A, Lin DL, Huisman TA. MRI with diffusion-weighted imaging in children and young adults with

simultaneous supra- and infratentorial manifestations of Sturge-Weber syndrome. *J Neuroradiol.* 2010;37(1): 51-59.

39. Basser PJ, Pierpaoli C. Microstructural and physiological features of tissues elucidated by quantitative-diffusion-tensor MRI. *J Magn Reson B.* 1996;111(3): 209-219.

40. Le Bihan D, Mangin JF, Poupon C, et al. Diffusion tensor imaging: concepts and applications. *J Magn Reson Imaging.* 2001;13(4): 534-546.

41. Moritani T, Kim J, Sato Y, Bonthius D, Smoker WR. Abnormal hypermyelination in a neonate with Sturge-Weber syndrome demonstrated on diffusion-tensor imaging. *J Magn Reson Imaging.* 2008;27(3): 617-620.

42. Alkonyi B, Chugani H, Behen M, et al. The role of the thalamus in neuro-cognitive dysfunction in early unilateral hemispheric injury: A multimodality imaging study of children with Sturge-Weber syndrome. *Eur J Paediatr Neurol.* 2010; in press.

43. Cho SH, Kim SH, Choi BY, et al. Motor outcome according to diffusion tensor tractography findings in the early stage of intracerebral hemorrhage. *Neurosci Lett.* 2007;421(2): 142-146.

44. Sivaswamy L, Rajamani K, Juhász C, Maqbool M, Makki M, Chugani HT. The corticospinal tract in Sturge-Weber syndrome: a diffusion tensor tractography study. *Brain Dev.* 2008;30(7): 447-453.

45. Liauw L, van Buchem MA, Spilt A, et al. MR angiography of the intracranial venous system. *Radiology.* 2000;214(3): 678-682.

46. Leach JL, Fortuna RB, Jones BV, Gaskill-Shipley MF. Imaging of cerebral venous thrombosis: current techniques, spectrum of findings, and diagnostic pitfalls. *Radiographics.* 2006;26(suppl 1): S19-41; discussion S42-43.

47. Vogl TJ, Stemmler J, Bergman C, Pfluger T, Egger E, Lissner J. MR and MR angiography of Sturge-Weber syndrome. *AJNR Am J Neuroradiol.* 1993;14(2): 417-425.

48. Cure JK, Holden KR, Van Tassel P. Progressive venous occlusion in a neonate with Sturge-Weber syndrome: demonstration with MR venography. *AJNR Am J Neuroradiol.* 1995;16(7): 1539-1542.

49. Juhász C, Chugani HT. An almost missed leptomeningeal angioma in Sturge-Weber syndrome. *Neurology.* 2007;68(3): 243.

50. Reiss AL, Abrams MT, Singer HS, Ross JL, Denckla MB. Brain development, gender and IQ in children. A volumetric imaging study. *Brain.* 1996;119(Pt 5): 1763-1774.

51. Kelley TM, Hatfield LA, Lin DD, Comi AM. Quantitative analysis

of cerebral cortical atrophy and correlation with clinical severity in unilateral Sturge-Weber syndrome. *J Child Neurol.* 2005;20(11): 867-870.

52. Juhász C, Lai C, Behen ME, et al. White matter volume as a major predictor of cognitive function in Sturge-Weber syndrome. *Arch Neurol.* 2007;64(8): 1169-1174.

53. Salter MG, Fern R. NMDA receptors are expressed in developing oligodendrocyte processes and mediate injury. *Nature.* 2005;438(7071): 1167-1171.

54. Birken DL, Oldendorf WH. N-acetyl-L-aspartic acid: a literature review of a compound prominent in 1H-NMR spectroscopic studies of brain. *Neurosci Biobehav Rev.* 1989;13(1): 23-31.

55. Pfund Z, Chugani DC, Juhász C, et al. Evidence for coupling between glucose metabolism and glutamate cycling using FDG PET and 1H magnetic resonance spectroscopy in patients with epilepsy. *J Cereb Blood Flow Metab.* 2000;20(5): 871-878.

56. McKnight TR. Proton magnetic resonance spectroscopic evaluation of brain tumor metabolism. *Semin Oncol.* 2004;31(5): 605-617.

57. Saunders DE. MR spectroscopy in stroke. *Br Med Bull.* 2000;56(2): 334-345.

58. Moore GJ, Slovis TL, Chugani HT. Proton magnetic resonance spectroscopy in children with Sturge-Weber syndrome. *J Child Neurol.* 1998;13(7): 332-335.

59. Sijens PE, Gieteling EW, Meiners LC, et al. Diffusion tensor imaging and magnetic resonance spectroscopy of the brain in a patient with Sturge-Weber syndrome. *Acta Radiol.* 2006;47(9): 972-976.

60. Batista CE, Chugani HT, Hu J, et al. Magnetic resonance spectroscopic imaging detects abnormalities in normal-appearing frontal lobe of patients with Sturge-Weber syndrome. *J Neuroimaging.* 2008;18(3): 306-313.

61. Stehling MK, Turner R, Mansfield P. Echo-planar imaging: magnetic resonance imaging in a fraction of a second. *Science.* 1991;254(5028): 43-50.

62. Leach JL, Holland SK. Functional MRI in children: clinical and research applications. *Pediatr Radiol.* 2010;40(1): 31-49.

63. Stapleton SR, Kiriakopoulos E, Mikulis D, et al. Combined utility of functional MRI, cortical mapping, and frameless stereotaxy in the resection of lesions in eloquent areas of brain in children. *Pediatr Neurosurg.* 1997;26(2): 68-82.

64. Brauer J, Neumann J, Friederici AD. Temporal dynamics of perisylvian activation during language processing in children and adults. *Neuroimage.*

2008;41(4): 1484-1492.

65. Byars AW, Holland SK, Strawsburg RH, et al. Practical aspects of conducting large-scale functional magnetic resonance imaging studies in children. *J Child Neurol.* 2002;17(12): 885-890.

66. O'Shaughnessy ES, Berl MM, Moore EN, Gaillard WD. Pediatric functional magnetic resonance imaging (fMRI): issues and applications. *J Child Neurol.* 2008;23(7): 791-801.

67. Müller RA, Chugani HT, Muzik O, Rothermel RD, Chakraborty PK. Language and motor functions activate calcified hemisphere in patients with Sturge-Weber syndrome: a positron emission tomography study. *J Child Neurol.* 1997;12(7): 431-437.

68. Altman NR, Bernal B. Brain activation in sedated children: auditory and visual functional MR imaging. *Radiology.* 2001;221(1): 56-63.

69. Bernal B, Altman N. Visual functional magnetic resonance imaging in patients with Sturge-Weber syndrome. *Pediatr Neurol.* 2004;31(1): 9-15.

70. Cohen L, Lehericy S, Henry C, et al. Learning to read without a left occipital lobe: right-hemispheric shift of visual word form area. *Ann Neurol.* 2004;56(6): 890-894.

71. Kikuchi S, Kubota F, Nishijima K, et al. Electroencephalogram-triggered functional magnetic resonance imaging in focal epilepsy. *Psychiatry Clin Neurosci.* 2004;58(3): 319-323.

72. Broyd SJ, Demanuele C, Debener S, Helps SK, James CJ, Sonuga-Barke EJ. Default-mode brain dysfunction in mental disorders: a systematic review. *Neurosci Biobehav Rev.* 2009;33(3): 279-296.

73. Minshew NJ, Keller TA. The nature of brain dysfunction in autism: functional brain imaging studies. *Curr Opin Neurol.* 2010;23(2): 124-130.

74. Bettus G, Guedj E, Joyeux F, et al. Decreased basal fMRI functional connectivity in epileptogenic networks and contralateral compensatory mechanisms. *Hum Brain Mapp.* 2009;30(5): 1580-1591.

75. Cole DM, Smith SM, Beckmann CF. Advances and pitfalls in the analysis and interpretation of resting-state FMRI data. *Front Syst Neurosci.* 2010;4:8.

76. Griffiths PD. Sturge-Weber syndrome revisited: the role of neuroradiology. *Neuropediatrics.* 1996;27(6): 284-294.

77. Griffiths PD, Boodram MB, Blaser S, et al. Abnormal ocular enhancement in Sturge-Weber syndrome: correlation of ocular MR and CT findings with clinical and intracranial imaging findings. *AJNR Am J Neuroradiol.* 1996;17(4): 749-754.

78. Chugani HT, Phelps ME, Mazziotta JC. Positron emission tomography study of human brain functional development. *Ann Neurol.* 1987;22(4): 487-497.

79. Kinnala A, Suhonen-Polvi H, Aarimaa T, et al. Cerebral metabolic rate for glucose during the first six months of life: an FDG positron emission tomography study. *Arch Dis Child Fetal Neonatal Ed.* 1996;74(3): F153-157.

80. Chugani HT. Metabolic Imaging: A Window on Brain Development and Plasticity. In: Waxman S, Becker C, Bunney W, et al, eds. *Neuroscientist.* Vol 5; 1999:29-40.

81. la Fougere C, Rominger A, Forster S, Geisler J, Bartenstein P. PET and SPECT in epilepsy: a critical review. *Epilepsy Behav.* 2009;15(1): 50-55.

82. O'Brien TJ, So EL, Mullan BP, et al. Subtraction ictal SPECT co-registered to MRI improves clinical usefulness of SPECT in localizing the surgical seizure focus. *Neurology.* 1998;50(2): 445-454.

83. Chugani HT, Mazziotta JC, Phelps ME. Sturge-Weber syndrome: a study of cerebral glucose utilization with positron emission tomography. *J Pediatr.* 1989;114(2): 244-253.

84. Lee JS, Asano E, Muzik O, et al. Sturge-Weber syndrome: correlation between clinical course and FDG PET findings. *Neurology.* 2001;57(2): 189-195.

85. Juhász C, Batista CE, Chugani DC, Muzik O, Chugani HT. Evolution of cortical metabolic abnormalities and their clinical correlates in Sturge-Weber syndrome. *Eur J Paediatr Neurol.* 2007;11(5): 277-284.

86. Benedek K, Juhász C, Chugani DC, Muzik O, Chugani HT. Longitudinal changes in cortical glucose hypometabolism in children with intractable epilepsy. *J Child Neurol.* 2006;21(1): 26-31.

87. Namer IJ, Battaglia F, Hirsch E, Constantinesco A, Marescaux C. Subtraction ictal SPECT co-registered to MRI (SISCOM) in Sturge-Weber syndrome. *Clin Nucl Med.* 2005;30(1): 39-40.

88. Comi AM. Sturge-Weber syndrome and epilepsy: an argument for aggressive seizure management in these patients. *Expert Rev Neurother.* 2007;7(8): 951-956.

89. Juhász C, Chugani H. Transient focal increase of interictal glucose metabolism in Sturge-Weber syndrome: Implications for epileptogenesis [abstract]. *Epilepsia.* 2009;50(suppl 11): 430.

90. Rischke R, Krieglstein J. Postischemic neuronal damage causes astroglial activation and increase in local cerebral glucose utilization of rat hippocampus. *J Cereb Blood Flow Metab.* 1991;11(1): 106-113.

91. Rischke R, Rami A, Bachmann U, Rabie A, Krieglstein J. Activated astrocytes, but not pyramidal cells, increase glucose utilization in rat hippocampal CA1 subfield after ischemia. *Pharmacology.* 1992;45(3): 142-153.

92. Tacconi MT. Neuronal death: is there a role for astrocytes? *Neurochem Res.* 1998;23(5): 759-765.

93. Batista CE, Chugani HT, Juhász C, Behen ME, Shankaran S. Transient hypermetabolism of the basal ganglia following perinatal hypoxia. *Pediatr Neurol.* 2007;36(5): 330-333.

94. Pu Y, Li QF, Zeng CM, et al. Increased detectability of alpha brain glutamate/glutamine in neonatal hypoxic-ischemic encephalopathy. *AJNR Am J Neuroradiol.* 2000;21(1): 203-212.

95. Pinton F, Chiron C, Enjolras O, Motte J, Syrota A, Dulac O. Early single photon emission computed tomography in Sturge-Weber syndrome. *J Neurol Neurosurg Psychiatry.* 1997;63(5): 616-621.

96. Rasmussen T, Villemure JG. Cerebral hemispherectomy for seizures with hemiplegia. *Cleve Clin J Med.* 1989;56(suppl Pt 1): S62-68; discussion S79-83.

97. Chugani HT, Muller RA, Chugani DC. Functional brain reorganization in children. *Brain Dev.* 1996;18(5): 347-356.

98. Muller RA, Rothermel RD, Behen ME, Muzik O, Mangner TJ, Chugani HT. Differential patterns of language and motor reorganization following early left hemisphere lesion: a PET study. *Arch Neurol.* 1998;55(8): 1113-1119.

99. Batista CE, Juhász C, Muzik O, Chugani DC, Chugani HT. Increased visual cortex glucose metabolism contralateral to angioma in children with Sturge-Weber syndrome. *Dev Med Child Neurol.* 2007;49(8): 567-573.

100. Chiron C, Raynaud C, Tzourio N, et al. Regional cerebral blood flow by SPECT imaging in Sturge-Weber disease: an aid for diagnosis. *J Neurol Neurosurg Psychiatry.* 1989;52(12): 1402-1409.

101. Griffiths PD, Boodram MB, Blaser S, Armstrong D, Gilday DL, Harwood-Nash D. 99mTechnetium HMPAO imaging in children with the Sturge-Weber syndrome: a study of nine cases with CT and MRI correlation. *Neuroradiology.* 1997;39(3): 219-224.

102. Reid DE, Maria BL, Drane WE, Quisling RG, Hoang KB. Central nervous system perfusion and metabolism abnormalities in Sturge-Weber syndrome. *J Child Neurol.* 1997;12(3): 218-222.

103. Bar-Sever Z, Connolly LP, Barnes PD, Treves ST. Technetium-99m-HMPAO SPECT in Sturge-Weber syndrome. *J Nucl Med.* 1996;37(1): 81-83.

104. Maria BL, Neufeld JA, Rosainz LC, et al. High prevalence of bihemispheric structural and functional defects in Sturge-Weber syndrome. *J Child Neurol.* 1998;13(12): 595-605.

105. Maria BL, Neufeld JA, Rosainz LC, et al. Central nervous system structure and function in Sturge-Weber syndrome: evidence of neurologic and radiologic progression. *J Child Neurol.* 1998;13(12): 606-618.

106. Aylett SE, Neville BG, Cross JH, Boyd S, Chong WK, Kirkham FJ. Sturge-Weber syndrome: cerebral haemodynamics during seizure activity. *Dev Med Child Neurol.* 1999;41(7): 480-485.

107. Riela AR, Stump DA, Roach ES, McLean WT, Jr., Garcia JC. Regional cerebral blood flow characteristics of the Sturge-Weber syndrome. *Pediatr Neurol.* 1985;1(2): 85-90.

108. Okudaira Y, Arai H, Sato K. Hemodynamic compromise as a factor in clinical progression of Sturge-Weber syndrome. *Childs Nerv Syst.* 1997;13(4): 214-219.

109. Yu TW, Liu HM, Lee WT. The correlation between motor impairment and cerebral blood flow in Sturge-Weber syndrome. *Eur J Paediatr Neurol.* 2007;11(2): 96-103.

110. Bilgin O, Vollmar C, Peraud A, la Fougere C, Beleza P, Noachtar S. Ictal SPECT in Sturge-Weber syndrome. *Epilepsy Res.* 2008;78(2-3): 240-243.

Neurosurgical Aspects of Sturge-Weber Syndrome

Ronald T. Grondin, M.D., M.Sc., F.R.C.S.C.

STURGE-WEBER SYNDROME IS A NEUROCUTANEOUS DISORDER characterized by facial port-wine nevus and leptomeningeal angiomatosis.[1] Sturge-Weber syndrome is associated with epilepsy in 75% to 90% of affected patients,[2-4] and is resistant to medical management in up to 60% of those patients.[2] Children with seizures beginning under a year of age have a worse prognosis for adequate seizure control.[5] These patients are also at greater risk of progressive neurologic deterioration and intellectual impairment.[4,6] Intellectual impairment is frequently seen in children with seizures due to Sturge-Weber syndrome.[4] Children whose seizures are controlled have improved intellectual funtion.[7] Although the optimum timing for surgical intervention in Sturge-Weber syndrome is not clear, it is felt that earlier intervention provides the greatest opportunity not only for intellectual development, but also for the recovery of deficits caused by surgery or the disease process.

Indications for surgical treatment of children with Sturge-Weber syndrome are similar to those for medically-resistant epilepsy from other causes.[5] Patients should have undergone a minimal course of maximally tolerated first-line anticonvulsant therapy. Patients who have failed medical management should then undergo appropriate diagnostic investigations to determine the location of seizure onset, including the extent of intracranial disease. These investigations include non-invasive electroencephalography (EEG) and video-EEG monitoring, neuropsychological evaluation, and neuroimaging (MRI, CT, SPECT, PET, and magnetoencephalography or MEG). It is important to understand the full extent of intracranial involvement prior to considering any surgical interventions, as at least 15% of patients will have bilateral brain lesions.[5,8]

Diagnostic Studies Related to Surgery

All patients with epilepsy should undergo routine, non-invasive EEG monitoring, including video-EEG monitoring. The site of EEG abnormalities should correspond to the area of the cortical lesion and seizure semiology. Failure to have convergent data should raise concern for the possibility of bilateral disease. Neuropsychological assessments may also contribute to identification of deficits in one hemisphere, and sometimes to a specific location.

Neuroimaging studies have particular value in Sturge-Weber syndrome because the lesion is usually readily identifiable on both CT and MRI imaging. The leptomeningeal angiomatosis is frequently associated with early calcification of cortical vessels, which is apparent on CT imaging, and may even be severe enough to be seen on plain skull radiographs. Functional MRI is a useful adjunct to identify areas of eloquent cortex however requires co-operation from the child and may not be feasible in all cases.

Other neuroimaging modalities are useful for identification of functional abnormalities of the brain. Ictal and inter-ictal SPECT are able to identify subtle differences in cortical blood flow. At the beginning of a seizure, there is an increase in blood flow to the area of seizure onset. Comparing ictal to interictal SPECT using commercially available subtraction software can aid in identifying the seizure onset zone. PET studies identify metabolic utilization of glucose by brain tissue between seizures; regions of hypometabolism show good correlation with the seizure onset zone. MEG is another neuroimaging modality that makes use of magnetic dipoles created by electrical discharges superimposed on anatomical MRI to localize seizure onset zones.[9] MEG can also be used to determine regions of eloquent cortex (motor, sensory, language). At our institution, we combine all available imaging studies using our intra-operative image guidance software in order to assist with the planning of a surgical resection **(Figure 7-1)**.

When there is a convergence of data indicating a regional area of seizure onset, then the likelihood of achieving seizure freedom by removing that abnormal area is greatly improved.

Surgical Management

The primary indication for neurosurgical intervention is medically resistant epilepsy. In addition to the neurologic manifestations of Sturge-Weber syndrome, these patients may also have other anomalies that require special consideration, including ocular abnormalities and angiomas of the trunk, face, and nasopharynx.[10] The anesthetic management of these patients should be

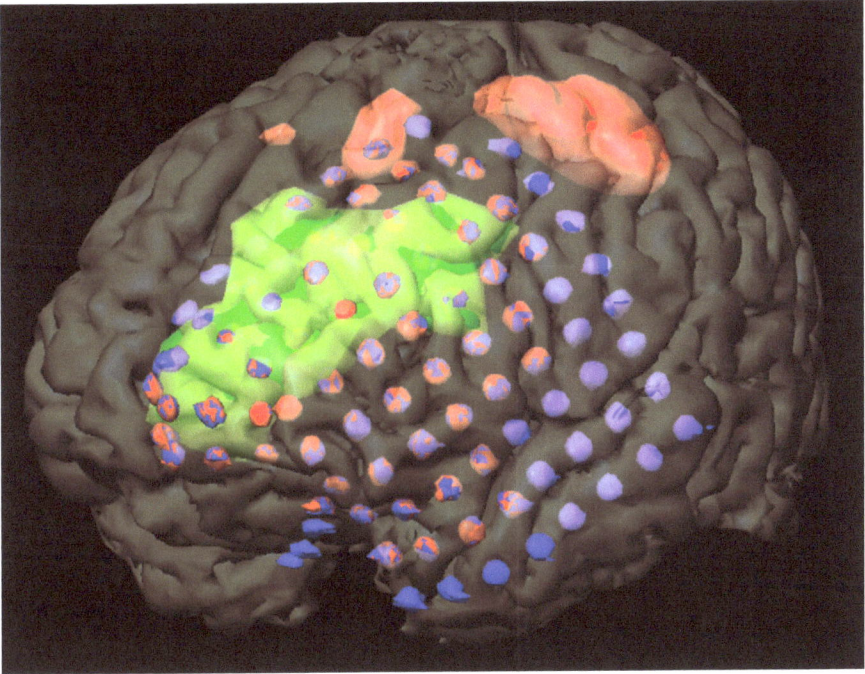

Figure 7-1. Three-dimensional reconstruction of data used for planning of epilepsy surgery. Subdural electrodes (blue) are shown superimposed over MRI reconstruction of cortical surface. Active EEG electrodes are shown in red. Ictal SPECT data (green) outlines area of increased blood flow at time of seizure onset. Functional MRI data acquired during a motor task (pink) identifies cortical areas involved in motor function.

planned in such a way as to avoid trauma to the hemangiomatous lesions and to prevent rise in intra-ocular and intracranial pressure.

In patients with medically resistant epilepsy, the goal of surgery is to remove or disconnect the entire lesion, in order to prevent further seizures. Depending on the extent of disease, this may be a focal cortical resection including the abnormal tissue, or may involve anatomic removal of the entire affected hemisphere.

When the lesion is unilateral and relatively localized, a focal cortical resection has a good likelihood of providing seizure freedom if the area of abnormality can be completely resected. The intracranial lesion may extend beyond what is visually identifiable, and the surgical resection is tailored to include at least the visible lesion **(Figure 7-2)**. Intra-operative electrocorticography is of assistance to ensure that no epileptogenic areas of cortex remain after resection of the lesion. In a retrospective review of 27 children undergoing epilepsy surgery for

Sturge-Weber syndrome, Bourgeois et al. reported seizure freedom in 89% of children in whom the lesion was completely resected, compared to only 30% in whom an incomplete resection was performed.[4] In this study, the completeness of resection was a statistically significant predictor of seizure freedom. The patients who underwent incomplete resection of the diseased area were older patients without motor deficits, and in whom attempts were made to preserve motor function.

In children with diffuse or extensive disease, complete disconnection of the affected hemisphere is required in order to achieve seizure freedom.[3-5, 11] Authors from earlier publications favored performing a complete anatomic hemispherectomy in order to achieve seizure freedom. However, the type of disconnection performed has not correlated to seizure freedom.[3, 11] Because of the risk of late complications from anatomic hemispherectomy, several authors have described functional hemispheric disconnection techniques that remove less tissue, while still achieving the desired goal of electrical disconnection of the hemisphere.[12-15] At least 15% of patients with the syndrome have bilateral disease.[16] If the disease is primarily affecting one hemisphere, then hemispheric disconnection may still be considered.[8]

When a focal lesion cannot be identified or safely resected, corpus callosotomy has been shown to be effective for certain seizure types, including tonic, atonic, and

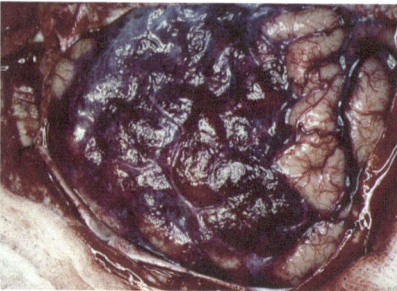

Figure 7-2. Intra-operative photographs demonstrating the vascular lesion in Sturge-Weber syndrome before (A) and after (B) focal cortical resection.

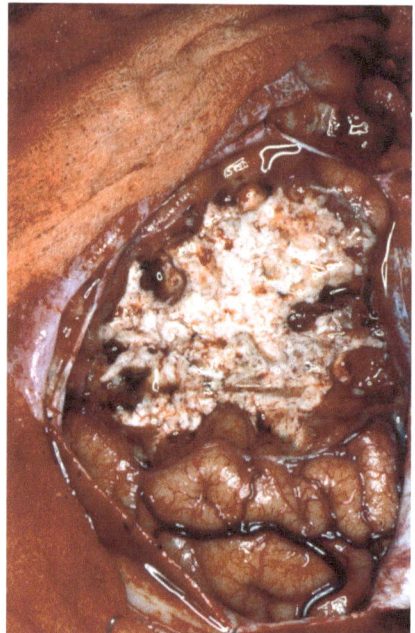

generalized tonic-clonic seizures.[8] Callosotomy has been used successfully in a limited number of patients with Sturge-Weber syndrome.[17] Callosotomy may eliminate some of the morbidity associated with these seizure types; it should be considered a palliative procedure when other surgical treatments are considered inappropriate, as complete seizure freedom is not often achieved. A vagus nerve stimulator is another palliative therapy that may be considered when neither focal resection nor hemispheric disconnection is not indicated.

Surgical Morbidity and Late Complications

Special consideration is needed for the surgical management of Sturge-Weber patients because of the abnormal vascularity of the affected tissue. Bleeding from the abnormal vessels is often difficult to control, and patients may experience significant blood loss during surgery. These issues may be addressed by meticulous attention to surgical hemostasis and by adequate volume replacement intra-operatively.[18]

The risks of neurologic deficit after resection depend on the cortical region involved. Focal resections involving primary motor cortex or hemispherectomy, either anatomic or functional, will result in immediate hemiparesis. Better recovery of motor function will be seen in younger patients. Kosoff and colleagues reported the outcomes of 32 patients treated by hemispherectomy worldwide, and noted that 47% of patients had complications in the immediate post-operative period.[3] These included bleeding, infection, severe headaches, persistent seizures requiring immediate re-operation, hydrocephalus, and hypertension. No deaths were reported.

Functional hemispherectomy techniques were first introduced in the 1970s in an attempt to avoid some of the known potential complications associated with anatomic hemispherectomy.[13-15] Superficial hemosiderosis is characterized by persistent intracranial bleeding with resultant hydrocephalus.[19] The development of hydrocephalus commonly occurred in patients undergoing anatomic hemispherectomy, and contributed to late morbidity. These late complications are less commonly encountered with newer disconnection techniques, in which resected tissue is minimized while still achieving electrical disconnection of the hemisphere.[15]

Outcomes

Post-operative measures of outcome include seizure freedom and developmental status. Overall, favorable seizure outcomes are achievable in a significant proportion of patients undergoing surgery. A large series of patients undergoing hemispherectomy (either anatomic or functional) for treatment of epilepsy secondary to Sturge-Weber syndrome demonstrated 81% seizure freedom,[3] regardless of the surgical technique used for disconnection.

Bourgeois and colleagues also reported good seizure outcomes in patients treated by either hemispheric disconnection or focal cortical resection for Sturge-Weber syndrome. In the setting of focal cortical resection, the completeness of resection of the lesion was a significant predictor of seizure freedom.[4] Furthermore, younger age at time of surgery was associated with better seizure outcomes. Improved developmental outcomes were also seen in children undergoing surgery at a younger age.[4] In this study complete resection of the lesion, or hemispherectomy, was also positively correlated with favorable developmental outcomes. Interestingly, no correlation was identified between duration of epilepsy prior to surgery and either seizure freedom or cognitive improvement.

In summary, children with epilepsy secondary to Sturge-Weber syndrome should be managed similarly to children with medically resistant epilepsy caused by other lesions. Surgery should be considered early in the course of medical management. Children who fail to respond to anticonvulsant therapy have a high likelihood of seizure freedom and improved developmental status following surgery.

REFERENCES

1. Bodensteiner JB, Roach ES. Sturge-Weber Syndrome: Introduction and Overview. In: Bodensteiner JB, Roach ES, eds. *Sturge Weber Syndrome*. Mt Freedom, NJ: Sturge Weber Foundation; 1999:1-10.

2. Di Rocco C, Tamburrini G. Sturge-Weber syndrome. *Childs Nerv Syst.* Aug 2006;22(8):909-921.

3. Kossoff EH, Buck C, Freeman JM. Outcomes of 32 hemispherectomies for Sturge-Weber syndrome worldwide. *Neurology.* Dec 10 2002;59(11):1735-1738.

4. Bourgeois M, Crimmins DW, de Oliveira RS, et al. Surgical treatment of epilepsy in Sturge-Weber syndrome in children. *J Neurosurg.* Jan 2007;106(1 Suppl):20-28.

5. Roach ES, Riela AR, Chugani HT, Shinnar S, Bodensteiner JB, Freeman J. Sturge-Weber syndrome: recommendations for surgery. *J Child Neurol.* Apr 1994;9(2):190-192.

6. Oakes WJ. The natural history of patients with the Sturge-Weber syndrome. *Pediatr Neurosurg.* 1992;18(5-6):287-290.

7. Ogunmekan AO, Hwang PA, Hoffman HJ. Sturge-Weber-Dimitri disease: role of hemispherectomy in prognosis. *Can J Neurol Sci.* Feb 1989;16(1):78-80.

8. Tuxhorn IE, Pannek HW. Epilepsy surgery in bilateral Sturge-Weber syndrome. *Pediatr Neurol.* May 2002;26(5):394-397.

9. Grondin R, Chuang S, Otsubo H, et al. The role of magnetoencephalography in pediatric epilepsy surgery. *Childs Nerv Syst.* Aug 2006;22(8):779-785.

10. Batra RK, Gulaya V, Madan R, Trikha A. Anaesthesia and the Sturge-Weber syndrome. *Can J Anaesth.* Feb 1994;41(2):133-136.

11. Arzimanoglou AA, Andermann F, Aicardi J, et al. Sturge-Weber syndrome: indications and results of surgery in 20 patients. *Neurology.* Nov 28 2000;55(10):1472-1479.

12. Kestle J, Connolly M, Cochrane D. Pediatric peri-insular hemispherotomy. *Pediatr Neurosurg.* Jan 2000;32(1):44-47.

13. Rasmussen T, Villemure JG. Cerebral hemispherectomy for seizures with hemiplegia. *Cleve Clin J Med.* 1989;56 Suppl Pt 1:S62-68; discussion S79-83.

14. Schramm J, Kral T, Clusmann H. Transsylvian keyhole functional hemispherectomy. *Neurosurgery.* Oct 2001;49(4):891-900; discussion 900-891.

15. Villemure JG, Daniel RT. Peri-insular hemispherotomy in paediatric epilepsy. *Childs Nerv Syst.* Aug 2006;22(8):967-981.

16. Bebin EM, Gomez MR. Prognosis in Sturge-Weber disease: comparison of unihemispheric and bihemispheric involvement. *J Child Neurol.* Jul 1988;3(3):181-184.

17. Rappaport ZH. Corpus callosum section in the treatment of intractable seizures in the Sturge-Weber syndrome. *Childs Nerv Syst.* Aug 1988;4(4):231-232.

18. Bruce DA. Neurosurgical Aspects of Sturge-Weber Syndrome. In: Bodensteiner JB, Roach ES, eds. *Sturge-Weber Syndrome.* Mt Freedom, NJ: Sturge-Weber Foundation; 1999:39-42.

19. Oppenheimer DR, Griffith HB. Persistent intracranial bleeding as a complication of hemispherectomy. *J Neurol Neurosurg Psychiatry.* Jun 1966;29(3):229-240.

Chapter 8

Mimics of Sturge-Weber Syndrome

Geoffrey Heyer, M.D.

IN MOST INSTANCES, THE DIAGNOSIS of Sturge Weber syndrome begins with recognition of the typical facial birthmark, the capillary malformation or port-wine stain. In order to avoid improper or unnecessary testing, the clinician must differentiate port-wine stains from other cutaneous vascular lesions that may share a similar appearance in the neonatal period. Unfortunately, historical inconsistencies in the classification of these birthmarks have led to a bewildering nomenclature and a great deal of diagnostic confusion. Terms such as *angioma* and *hemangioma* have been used generically to describe a variety of vascular lesions with different etiologies and clinical manifestations. Hybrid terms such as "capillary-cavernous hemangioma" further add to the confusion.

In 1982, Mulliken and Glowacki recognized that many of the entities referred to as hemangiomas were actually structural malformations of capillaries, veins, arteries, or lymphatic vessels, or a combination of vessel types. They proposed that vascular birthmarks be classified as either *malformations* or *hemangiomas* based on their cellular features in relation to their natural history and clinical course.[1] Vascular malformations are present at birth, grow commensurately with the child, and never regress. In contrast, hemangiomas may or may not be apparent at birth, and they exhibit rapid proliferation followed by slow involution. At the cellular level, hemangiomas are further distinguished from vascular malformations by increased endothelial cell activity during the proliferating phase. The vascular channels of malformations have normal endothelial cell turnover without hypercellularity.

The port-wine stain of Sturge Weber syndrome belongs to the *vascular malformation* category. The birthmark consists of venule-like channels located within the dermis. In keeping with the classification scheme above, port-wine stains are always present at birth and grow at the same pace as the child. They may expand spontaneously during puberty or pregnancy as a result of hormonal modulation, or develop nodular enlargements from arterio-venous shunting or thrombosis following trauma. However, enlargement of a vascular malformation in older children and adults should not be confused with the early proliferation of a hemangioma. The facial distribution of the port-wine stain determines the risk of Sturge Weber syndrome. Enjolras and colleagues found that ocular and brain abnormalities occurred exclusively in patients with upper facial involvement, with or without more widespread lesions.[2] Among 310 patients described by Tallman and colleagues, all those with extracutaneous findings of Sturge Weber syndrome had stain involvement of the eyelids: 91% had upper and lower eyelid involvement and 9% had only lower lid lesions.[3] Port-wine stains that affected the face bilaterally or had large unilateral facial distributions were associated with a significantly higher likelihood of eye and brain complications. Accordingly, Sturge Weber syndrome should be suspected in any individual with unilateral or bilateral periocular capillary malformations.

Red birthmarks are common. The remainder of this chapter focuses on some of the vascular lesions and neurocutaneous disorders that can be confused with Sturge Weber syndrome. Syndromes with vascular birthmarks that characteristically affect the extremities (e.g., Klippel-Trenaunay syndrome and Parkes Weber syndrome) will not be discussed.

Hemangiomas and PHACE Syndrome

Hemangiomas are common, benign vascular tumors affecting 2.5% to 10% of infants. They occur more commonly in females, Caucasians, and premature infants. Girls are affected more than boys at an approximate ratio of 3:1. The female predilection becomes more exaggerated in PHACE syndrome (described below) as the ratio increases to 7-9:1. The tumors are clinically heterogeneous, with variations in appearance based on depth, location, and stage of evolution. A hemangioma may or may not be evident at birth. Early lesions may be clinically subtle, emerging as a pallid macule surrounded by telangiectasias. In early infancy, tumor proliferation begins. With growth, the superficial lesions become recognizable as bright red, non-compressible plaques or nodules (**Figure 8-1**). Regardless of depth or location, hemangiomas reach 80% of their final size by approximately three months of age.[4] The early proliferative stage ends around five months, at which time most hemangioma growth is complete. A period of slow, spontaneous involution follows for an average of 2 to 6 years.[5] A rarer type of hemangioma appears as a reticulated or telangiectatic (**Figure 8-2**).

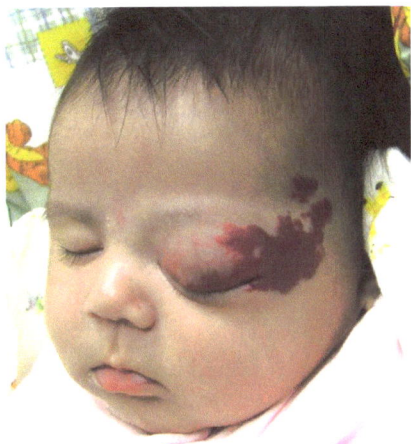

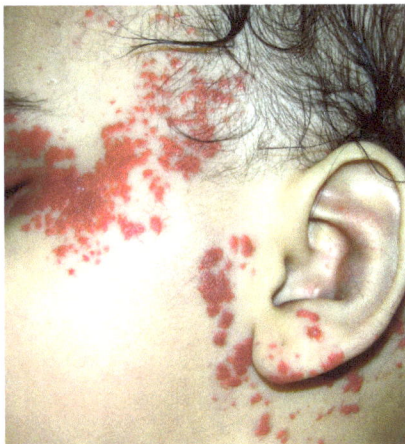

Figure 8-1A and B. A. Plaque-like facial hemangioma in the early progressive phase. This patient had multiple extracutaneous anomalies consistent with PHACE syndrome. B. A second child with PHACE syndrome illustrating the variable appearance of the facial lesion.

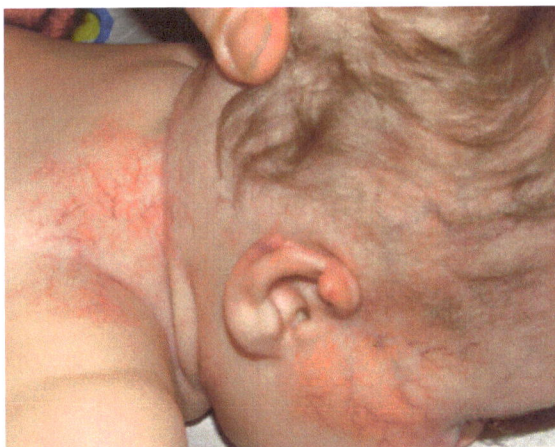

Figure 8-2. Reticulated or telangiectatic appearing hemangioma. This child had multiple associated cerebral vascular anomalies consistent with PHACE syndrome. Also, note necrosis of the ear lobe. *Reproduced from Heyer, et al[7]*

Some patients with facial hemangiomas exhibit additional associated structural anomalies of the brain, cerebral vasculature, eyes, aorta, and/or heart in a neurocutaneous disorder of unknown etiology called PHACE syndrome. The PHACE acronym refers to malformations of the *p*osterior fossa, facial *h*emangiomas, *a*rterial cerebral vascular anomalies, *c*ardiovascular anomalies, and abnormalities of the *e*ye. When ventral developmental defects such as *s*ternal clefting or *s*upraumbilical raphe are present, the PHACES acronym may be used. The disorder represents a spectrum of disease with most patients exhibiting only one extracutaneous feature. A consensus statement published in 2009 differentiates criteria as *major* and *minor*.[6] A definite PHACE syndrome diagnosis requires the presence of a characteristic facial hemangioma plus one major criterion or two minor criteria. The consensus group also suggested that this disorder might occur with a hemangioma affecting the neck, chest, or arm only, or with no detectable hemangioma. Until further study, these diagnoses should be referred to as *possible* PHACE syndrome.

The hallmark of this neurocutaneous disorder is the plaque-like and segmental facial hemangioma. These vascular lesions tend to correspond to one or more of the facial segments described by Haggstrom and colleagues.[7] During the early phase of proliferation, the hemangioma may be mistaken for a port-wine stain. However, the birthmarks become easily distinguishable as the hemangioma evolves.

Abnormal neurological signs and symptoms develop as consequence of structural brain malformations and cerebral vascular anomalies. Multiple brain lesions have been reported including a spectrum of cerebellar defects and posterior fossa cysts as well as cortical dysplasia. The vasculopathy of PHACE syndrome affects cerebral arteries and can be divided into two categories: congenital anomalies and progressive stenoses and occlusions. The progressive arterial disease leads to ischemic stroke and moyamoya syndrome in some affected children. Among 115 reported PHACE patients, 77% had congenital and/or progressive cerebral vascular anomalies.[8] Central nervous system hemangiomas may be detected by MRI, but they rarely cause neurologic signs or symptoms. Additionally, areas of abnormal gadolinium-enhancement are often found overlying regions of cerebral or cerebellar dysplasia. These may represent small-vessel anomalies or central nervous system hemangiomas.

Less commonly identified complications of PHACE syndrome include abnormalities of the heart and eye and ventral developmental defects. Several cardiac anomalies involving the heart itself are recognized, but aortic coarctation has been reported most frequently. The aortic coarctations of PHACE syndrome often have associated hypoplasia of the aortic arch, adding substantial risk during surgical repair.[9] Approximately 20% to 25% of PHACE patients have unusual ophthalmologic findings. These include microphthalmia, choroidal hemangiomas, strabismus, congenital cataracts, glaucoma, retinal colobomas, and the morning glory disc anomaly. Ventral developmental defects manifest as sternal clefting and supraumbilical raphe. Sternal clefting ranges in severity from a cutaneous sternal pit to complete separation of the sternal bars.

The presence of a characteristic infantile hemangioma warrants clinical evaluation by neurology, cardiology, dermatology, and ophthalmology. Gadolinium-enhanced MRI with MR angiography should be performed to assess the brain and cerebral vasculature. These individuals should also undergo echocardiography. Older children and adults found to have cerebral vascular anomalies should be questioned about resolved facial hemangiomas, as other PHACE anomalies may be present.

Macrocephaly-Capillary Malformation Syndrome

Macrocephaly-capillary malformation (M-CM) syndrome is a rare neurocutaneous disorder of unknown etiology. Characteristic features include macrocephaly (head circumference >95th percentile), facial and limb asymmetry, somatic overgrowth, developmental delay, and capillary malformations (port-wine stains) affecting the torso, extremities, and central face. The port-wine stains have a confluent or reticulated pattern. Authors in the past had incorrectly referred to the reticulated capillary malformations as "cutis marmorata telangiectatica congenita," naming the disorder *macrocephaly-cutis marmorata telangiectatica congenita syndrome* (M-CMTC). Toriello and Mulliken proposed renaming the disorder to M-CM to accurately reflect the cutaneous vascular lesions.[10]

Approximately half of the patients with M-CM have a facial port-wine stain.[11] Unlike Sturge Weber syndrome, the birthmark more commonly affects central facial structures, including the lip, philtrum, nose, and glabella. Moreover, the color of the birthmark may fade with age in some M-CM patients.[12] Macrocephaly is typically congenital and may be progressive. Clinical signs suggestive of increased intracranial pressure warrant immediate clinical and imaging evaluations. Conway and colleagues found several radiographic abnormalities in M-CM patients, including white matter irregularities, rapid brain growth, ventriculomegaly (frequently obstructive), and herniation of the cerebellar tonsils.[13] They suggest that rapid growth of the cerebellum leads to mechanical compromise of the posterior fossa and acquired tonsillar herniation. Mild to moderate developmental delay is a common finding among M-CM patients.

Nova Syndrome

In 1979, Harvey Nova described a maternal grandmother, mother, and three male children with central facial port-wine stains and mega cisterna magna.[14] The proband and one brother had hydrocephalus with agenesis of the cerebellar vermis. Head circumferences for the other three family members were not reported. This disorder has been referred to as Nova syndrome. Similar case descriptions have been published since.

Wyburn-Mason Syndrome

The clinical features of Wyburn-Mason syndrome (Bonnet-Dechaume-Blanc syndrome) comprise ipsilateral arteriovenous malformations (AVM) of the brain and retina and a congenital facial stain. Bonnet and coauthors described features of the syndrome among French patients in 1937.[15] In 1943, Wyburn-Mason published a case series in the English-language literature.[16] The etiology of the disorder remains unknown.

The facial birthmark has been described as a capillary malformation but, rather, represents a quiescent Schobinger stage-1 AVM.[17,18] Facial vascular lesions are inconsistent in this syndrome and are not required for the diagnosis. When they occur, they tend to be unilateral or midfacial. However, bilateral lesions have been reported. Retinal AVMs affect 25% to 30% of patients.[16,19] They appear as irregular, tortuous, and ectatic vessels with minimal observable differences between arteries and veins. The locations of cerebral AVMs vary. Common sites include regions along the visual pathway: the orbit, optic nerve, optic chiasm, thalamus, hypothalamus, and suprasellar area. Clinical symptoms in Wyburn-Mason syndrome relate to the location and size of each cerebral AVM and to the possibility of hemorrhage. Retinal lesions are typically stable, although spontaneous hemorrhage leading to visual impairment has been described.

Capillary Malformation-AVM (CM-AVM) Syndrome

Capillary malformation-AVM (CM-AVM) syndrome is an autosomal disorder caused by mutation of the *RASA1* gene.[20] Multifocal cutaneous capillary malformations constitute the cardinal feature of CM-AVM. These birthmarks range in size and may occur anywhere on the body, including the face. Revencu and colleagues found that 16 patients from 13 families with a clinical diagnosis of Parkes Weber syndrome had *RASA1* mutations.[21] They also found intracranial and extracranial vascular anomalies associated in patients with multifocal capillary malformations and gene mutation. The intracranial vascular anomalies included vein of Galen aneurismal malformations, cerebral and meningeal AVMs, and arteriovenous fistulae. AVMs occurred in the frontal lobes and in the posterior fossa. Notably, patients with *RASA1* mutations did not undergo routine MRI screening, so the scope and frequency of intracranial pathology has not yet been determined.

REFERENCES

1. Mulliken MD, Glowacki J. Hemangiomas and vascular malformations in infants and children; A classification based on endothelial characteristics. *Plast Reconstr Surg.* 1982; 69: 412-20.

2. Enjolras O, Riche MC, Merland JJ. Facial port-wine stains and Sturge-Weber syndrome. *Pediatrics* 1985;76:48-51.

3. Tallman B, Tan OT, Morelli JG et al. Location of port-wine stains and the likelihood of ophthalmic and/or central nervous system complications. *Pediatrics* 1991;87:323-327.

4. Chang LC, Haggstrom AN, Drolet BA et al. Growth characteristics of infantile hemangiomas: implications for management. *Pediatrics* 2008;122:360-367.

5. Metry DW, Hebert AA. Benign cutaneous vascular tumors of infancy: when to worry, what to do. *Arch Dermatol* 2000;136:905-914.

6. Metry D, Heyer G, Hess C et al. PHACE Syndrome Research Conference Consensus Statement on Diagnostic Criteria for PHACE Syndrome. *Pediatrics* 2009;124:1447-1456.

7. Haggstrom AN, Lammer EJ, Schneider RA, Marcucio R, Frieden IJ. Patterns of infantile hemangiomas: new clues to hemangioma pathogenesis and embryonic facial development. *Pediatrics* 2006;117:698-703.

8. Heyer GL, Dowling MM, Licht DJ et al. The cerebral vasculopathy of PHACES syndrome. *Stroke* 2008;39:308-316.

9. Bijulal S, Sivasankaran S, Krishnamoorthy KM, Titus T, Tharakan JA, Krishnamanohar SR. Unusual coarctation-the PHACE syndrome: report of three cases. *Congenit Heart Dis* 2008;3:205-208.

10. Toriello HV, Mulliken JB. Accurately renaming macrocephaly-cutis marmorata telangiectatica congenita (M-CMTC) as macrocephaly-capillary malformation (M-CM). *Am J Med Genet A* 2007;143A:3009.

11. Wright DR, Frieden IJ, Orlow SJ et al. The misnomer "macrocephaly-cutis marmorata telangiectatica congenita syndrome": report of 12 new cases and support for revising the name to macrocephaly-capillary malformations. *Arch Dermatol* 2009;145:287-293.

12. Amitai DB, Fichman S, Merlob P, Morad Y, Lapidoth M, Metzker A. Cutis marmorata telangiectatica congenita: clinical findings in 85 patients. *Pediatr Dermatol* 2000;17:100-104.

13. Conway RL, Pressman BD, Dobyns WB et al. Neuroimaging findings in macrocephaly-capillary malformation: a longitudinal study of 17 patients. *Am J Med Genet A* 2007;143A:2981-3008.

14. Nova HR. Familial communicating hydrocephalus, posterior cerebellar agenesis, mega cisterna magna, and port-wine nevi. Report on five members of one family. *J Neurosurg* 1979;51:862-865.

15. Bonnet P, Dechaume J, Blanc E. L'anéurysme cirsöide de la rétine (anéurysme recémeux): ses relations avec l'anéurysme cirsöide de la face et l'anéurysme cirsöide du cerveau. *J Med Lyon* 1937;18:165-178.

16. Wyburn-Mason R. Arteriovenous aneurysm of mid-brain and retina, facial naevi and mental changes. *Brain* 1943;66:12-203.

17. Theron J, Newton TH, Hoyt WF. Unilateral retinocephalic vascular malformations. *Neurorad* 1974;7:185-196.

18. Patel U, Gupta SC. Wyburn-Mason syndrome. A case report and review of the literature. *Neurorad* 1990;31:544-546.

19. Dayani PN, Sadun AA. A case report of Wyburn-Mason syndrome and review of the literature. *Neurorad* 2007;49:445-456.

20. Eerola I, Boon LM, Mulliken JB et al. Capillary malformation-arteriovenous malformation, a new clinical and genetic disorder caused by RASA1 mutations. *Am J Hum Genet* 2003;73:1240-1249.

21. Revencu N, Boon LM, Mulliken JB et al. Parkes Weber syndrome, vein of Galen aneurysmal malformation, and other fast-flow vascular anomalies are caused by RASA1 mutations. *Hum Mutat* 2008;29:959-965.

Chapter 9

The Sturge-Weber Foundation

Karen L. Ball, President and CEO

THE STURGE-WEBER FOUNDATION (SWF) incorporated in our unfinished basement in 1987 as a 501(c) 3 not-for-profit organization for individuals with Sturge-Weber syndrome and their families, health care professionals, and others concerned with the syndrome. It was conceived in a daring dream to improve quality of life not only for our daughter, Kaelin, who was diagnosed at birth, but also for those living with the syndrome. We knew one day, with collaboration and tenacity, the SWF could provide families more than hope, but answers through research. Since its incorporation, the Foundation has maintained a registry of individuals with Sturge-Weber syndrome, Klippel-Trenaunay and port wine birthmarks (in 1993, the board of directors voted to expand the foundation's mission to include these individuals). As of September 2010, the registry has grown to 3,221 active members with Sturge-Weber syndrome and 1,128 other known affected individuals with a birthmark or KT.

Karen Ball and her daughter, Kaelin.

The SWF mission is to educate the public, empower families and individuals, and instigate research into the causes of these birthmarks and syndromes. Today the Foundation and its members also partner with caregivers, scientific researchers, and educators dedicated to changing the natural history of the syndrome. The Foundation is honored to count 5,933 individuals actively engaged in this endeavor.

The Sturge-Weber Foundation provides information and services to members, the medical professionals who care for its members, and the general public, which includes various governmental agencies. The Foundation publishes a periodic newsmagazine, brochures, and relevant educational support materials. The Foundation also has a comprehensive website and utilizes multimedia presentations to address diverse topics that relate to living with those affected and to managing their care. The Foundation's 23-year-old, international natural history registry provides physicians with critical data on how to manage the course of the syndrome and identifies emerging trends in co-morbidities. The registry will enable better therapeutic options to manage and correct these trends with further research.

In addition, the Foundation provides one-on-one counseling, which has become a crucial lifeline for all those affected by these syndromes. The Foundation sponsors international and regional family conferences and fosters local family days to build friendships and family networks. It offers an array of educational programs for medical and scientific professionals, such as workshops, exhibits, and sponsorship of key National Institutes of Health conferences. This network of influence and awareness has increased the pace of discovery, but there is still much more to be accomplished!

THE ORGANIZATION

The Sturge-Weber Foundation maintains an office staffed by compassionate and knowledgeable individuals ready to assist and refer families as needed. The staff guides dedicated families who generate local awareness during the annual Month of Awareness and host local fundraising events. They also collaborate with diverse volunteers, the Centers of Excellence, and scientific researchers.

Volunteers

Board of Directors is responsible for setting policy and providing direction to achieve the Foundation's mission and goals.

Medical Advisory Board is made up of physicians, specialists with experience concerning the treatment of patients with these syndromes, and scientists dedicated to investigating the pathogenesis of the syndromes and effective therapeutic options.

Area Representatives work at local, national, and international levels to maintain contacts with families in their respective regions, host fundraising events, and educate their local communities and healthcare personnel.

Centers of Excellence

As of April 2010, The SWF has established 10 Centers of Excellence in cities throughout the USA. These centers provide the comprehensive care necessary for treating adults and children who have a port wine (PW) birthmark, Sturge-Weber syndrome (SWS), or Klippel-Trenaunay (KT) syndrome. Each center is staffed by a team of specialists who collaborate in the evaluation and management of each patient. This team approach ensures the individual's treatment plan is carefully developed and coordinated.

THE CATALYST

The Sturge-Weber Foundation has been a catalyst for change since its incorporation. The Foundation quickly gained top clinical and scientific advisors. With each new family calling for assistance, and with each researcher offering his or her expertise, the Foundation gets one step closer to discovering the causes of Sturge-Weber syndrome and other birthmarks. The informal information exchange developed into a more formalized Natural History Registry. It has expanded over the years as more information became available into the natural course of these syndromes.

A turning point for the Foundation came in 1999, when the National Institutes of Health funded a Sturge-Weber symposium. The Foundation co-hosted this meeting and changed the course of scientific research by providing a reliable source of clinical data and seed grant funding. This data has been cited in numerous medical journal articles on Sturge-Weber syndrome and presented at many medical conferences either as an abstract or during platform sessions. The data has also documented emerging co-morbidities and helped investigators obtain further research funding from the National Institutes of Health.

While the research investigations increased, the Foundation was still frustrated at the pace of discovery on the syndrome. The Foundation hosted a translational research summit in 2006. This meeting brought together leading experts in the field and the National Institutes of Health to discuss development of a strategic research plan and to identify translational research opportunities. There has been a renewed spike of interest and the development of a related mouse model, which we hope will spur new therapies for treatment.

The Foundation participates in coalitions and umbrella organizations such as the National Organization for Rare Disorders (NORD), the American Brain Coalition (ABC), the Coalition of Skin Diseases (CSD), and governmental entities such as the Office of Rare Diseases Research (ORDR). This type of networking fosters immediate and future Foundation member benefits at the local and federal level. Together we are a dynamic catalyst of tenacity and hope.

The Foundation has successfully proven that comprehensive global education, outreach, and awareness can change the course of a disease and the lives of those affected. It has engaged the brightest minds and most dedicated volunteers to increase the pace of discovery. The Foundation continues to make significant impact on the quality of life of its members and those affected. By being open to collaboration and new ways of thinking about the disease state, the Foundation improves the delivery of patient care. It hosts seminars to open dialogue and promote translational studies as well as to identify synergies amongst diseases. This format has transformed the attitude of researchers from ignoring our rare disease to inquiring first about our findings and then applying for Foundation research grants.

In 2010, the Foundation remains the leading international not-for-profit organization that provides Sturge-Weber syndrome clinical data, critical tissue donations, and member natural history data to research scientists. The Foundation continues to strive for the development of new resources and new technologies to serve our members and scientists across the world. This is accomplished with diligent and targeted networking, organizational and research strategic plans, and research seed grant awards. The Foundation has begun a dialogue with bio-tech companies to educate them on the syndrome and the co-morbidities, and to encourage them to invest their resources in the Foundation's endeavors and membership.

The Foundation's families still hold Sturge-Weber syndrome-related secrets ready to be discovered. The Foundation is a rich storehouse of genealogical and medical history data waiting for more families and investigators to join forces with us. Dreams really do come true with a strong belief, unbridled hope, and unending love for those who live courageously each day with Sturge-Weber syndrome.

Organizations and Websites

The following entries reflect select information available in April 2010, when this appendix was prepared. These entries are meant as a starting point for further education. However, addresses, telephone numbers, and online information are dynamic and thus some information will not remain current. The fastest way to get information tailor fit to your needs is to contact the Sturge-Weber Foundation directly, through swf@sturge-weber.org or 1-800-627-5482. The website, www.sturge-weber.org, also has a more comprehensive listing of resources.

American Council of the Blind
2200 Wilson Boulevard
Suite 650
Arlington, Va 22201
(800) 424-8666
(202) 467-5081
(703) 465-5085 Fax
www.acb.org

The American Council of the Blind (ABC) is a national not-for profit advocacy organization for people who are blind or visually impaired. They produce various publications, including the "Braille Forum."

Produced in Braille, large print, multi-media and computer software, it contains articles on employment, legislation, sports and leisure activities, new products and services, human interest stories, and other information of interest to blind and visually impaired people.

The Arc of the United States
1660 L Street, NW Suite 301
Washington, DC 20036
(800) 433-5255
(202) 534-3700
(202) 534-3731
www.thearc.org

The Arc is the world's largest community-based organization of and for people with intellectual and developmental disabilities. The Arc works on the national, state, and local levels to provide services, guidance, advocacy, free public educational opportunities, and resources to parents and other individuals, organizations, and communities concerned with solving the problems caused by developmental disabilities.

Family Resource Center on Disabilities
(Formerly Coordinating Council for Handicapped Children)
20 East Jackson Blvd., Room 300
Chicago, IL 60604
(800) 952-4199
(312) 939-3513
(312) 939-3519 (TDD)
(312) 939-7297 Fax
www.FRCD.org

This parent and professional organization informs and activates parents regarding the special-education rights of their children with handicaps.

Learning Disabilities Association of America
4156 Library Road
Pittsburgh, PA 15234
(412) 341-1515
(412) 344-0224 Fax
www.ldanatl.org

This organization is devoted to the well-being and education of children and adults with learning disabilities and will respond to request for information on any related topic.

Lighthouse International

111 East 59th St.

New York, NY 10022

(800) 829-0500

www.lighthouse.org

The Lighthouse International is dedicated to fighting vision loss through prevention, treatment, and empowerment.

National Organization for Rare Disorders

55 Kenosia Avenue

P.O. Box 1968

Danbury, CT 06813

(800) 999-6673 (voice mail only)

(203) 744-0100

(203) 797-9590 (TDD)

(203) 798-2291

www.rarediseases.org

The National Organization for Rare Disorders is a not-for-profit federation of voluntary health organizations dedicated to helping people with rare, "orphan" diseases and assisting the organizations that serve them. NORD is committed to the identification, treatment, and cure of rare disorders through programs of education, advocacy, research, and service. Its primary goal is to educate the public and the medical community by serving as an international clearinghouse of information on rare" diseases. NORD is also committed to increasing funding for clinical research for rare disorders. In addition NORD links thousands of families with others who have the same disorders through its Networking Program, gives referrals to additional organizations, clearinghouses, registries, and support groups that benefit families with rare disorders, and offers Medication Assistance Programs that provide certain prescription pharmaceuticals to needy patients. NORD sponsors an annual Patient/Family conference and an annual conference for support group leaders. NORD also complies the medical textbook, *Physicians' Guide to Rare Diseases*.

Office of Rare Diseases Research
National Institutes of Health
6100 Executive Boulevard
Room 3A07, MSC 7518
Bethesda, Maryland 20892-7518
(301) 402-4336
(301) 480-9655 (fax)
ordr@od.nih.gov

The ORDR has been an integral partner with patient advocacy organizations, investigators, and institutions to drive the clinical and scientific collaborations that improve the quality of life and care for those individuals living with a rare disease. The ORDR is effective in working with diverse stakeholders and the National Institutes of Health to facilitate partnerships and host workshops that stimulate change. The ORDR website has links to a broad range of information about rare diseases including definitions, causes, treatments, and publications, as well as information on rare diseases research and research resources, patient support groups, and genetic testing laboratories and clinics.

Featured Links on the Home Page:

Rare Diseases and Related Terms

A searchable list of almost 7,000 rare diseases with links to information from federally supported databases and information sources.

NLM Gateway

Allows users to search across multiple resources offered on the National Library of Medicine's website.

PubMed

Access to more than 16 million MEDLINE citations, life science journals, and links to many sites providing full text articles.

Office of Orphan Products Development, FDA

Promotes the development of products that demonstrate promise for the diagnosis and/or treatment of rare diseases or conditions.

Talking Glossary of Genetics

The National Human Genome Research Institute (NHGRI) created the

Talking Glossary of Genetic Terms to help everyone understand the terms and concepts used in genetic research. In addition to definitions, specialists in the field of genetics share their descriptions of terms, and many terms include images, animation, and links to related terms.

Positive Exposure

Rick Guidotti
43 East 20th Street
New York, NY 10003
(212) 420-1931
(212) 228-0592 Fax
rick@positiveexposure.org

Positive Exposure, founded in 1997 by former fashion photographer Rick Guidotti and Diane McLean, M.D., Ph.D., MPH, is a highly innovative arts organization working with individuals living with genetic difference. Through vigorous cross-sector partnerships with health advocacy organizations, governmental agencies, and educational institutions, Positive Exposure utilizes the visual arts to significantly impact the fields of genetics, mental health, and human rights.

Their programs support and promote human dignity through Positive Exposure's Spirit of Difference photographic image data bank and video interviews of persons, particularly children, living with genetic conditions.

The Sturge-Weber Foundation

P.O. Box 418
Mt. Freedom, NJ 07970
(800) 627-5482
(973) 895-4445
(973) 895-4846 Fax
www.sturge-weber.org
swf@sturge-weber.org

The Sturge-Weber Foundation is a not-for-profit organization for parents, professionals, and others concerned with Sturge-Weber syndrome. The Sturge-Weber Foundation acts as a clearinghouse for information on all aspects of Sturge-Weber syndrome and offers support to all interested parties. The group seeks to generate awareness and to educate the medical community, government agencies, and the general public about the syndrome, to promote the funding of ongoing medical research on the disorder, and to maintain a natural history registry of individuals with Sturge-Weber syndrome, Klippel-Trenaunay, and birthmarks. The Sturge-Weber Foundation provides referrals and offers a variety of educational and support materials including a medical textbook, resource guide, and brochures on all aspects of Sturge-Weber syndrome. Other educational resources available from the foundation include a website, social media outlets, multimedia presentations, member conferences, and a periodic newsmagazine which can be downloaded from the website.

Index

www.ingramcontent.com/pod-product-compliance
Lightning Source LLC
Chambersburg PA
CBHW041757190326
41458CB00041B/6529/J